OPTICAL COMMUNICATION THEORY AND TECHNIQUES

OPTICAL COMMUNICATION THEORY AND TECHNIQUES

Edited by

ENRICO FORESTIERI
Scuola Superiore Sant'Anna, Pisa, Italy

 Springer

Enrico Forestieri
Centro Eccellenza per Ingegneria Reti di Comunicazione
Sculoa Superiore S. Anna
145/147 Via Cisanello
Pisa 56124
Italy

Library of Congress Cataloging-in-Publication Data

A C.I.P. Catalogue record for this book is available from the Library of Congress.

OPTICAL COMMUNICATION THEORY AND TECHNIQUES/ edited by Enrico
Forestieri
    p.cm.
    Includes bibliographical references and index.

ISBN 978-1-4419-3577-9   e-ISBN 978-0-387-23136-5   Printed on acid-free paper.

Printed in the United States of America.

9 8 7 6 5 4 3 2 1

springeronline.com

Contents

Contents

Preface

Since the advent of optical communications, a great technological effort has been devoted to the exploitation of the huge bandwidth of optical fibers. Starting from a few Mb/s single channel systems, a fast and constant technological development has led to the actual 10 Gb/s per channel dense wavelength division multiplexing (DWDM) systems, with dozens of channels on a single fiber. Transmitters and receivers are now ready for 40 Gb/s, whereas hundreds of channels can be simultaneously amplified by optical amplifiers.

Nevertheless, despite such a pace in technological progress, optical communications are still in a primitive stage if compared, for instance, to radio communications: the widely spread on-off keying (OOK) modulation format is equivalent to the rough amplitude modulation (AM) format, whereas the DWDM technique is nothing more than the optical version of the frequency division multiplexing (FDM) technique. Moreover, adaptive equalization, channel coding or maximum likelihood detection are still considered something "exotic" in the optical world. This is mainly due to the favourable characteristics of the fiber optic channel (large bandwidth, low attenuation, channel stability, ...), which so far allowed us to use very simple transmission and detection techniques.

But now we are slightly moving toward the physical limits of the fiber and, as it was the case for radio communications, more sophisticated techniques will be needed to increase the spectral efficiency and counteract the transmission impairments. At the same time, the evolution of the *techniques* should be supported, or better preceded, by an analogous evolution of the *theory*. Looking at the literature, contradictions are not unlikely to be found among different theoretical works, and a lack of standards and common theoretical basis can be observed. As an example, the performance of an optical system is often given in terms of different, and sometimes misleading, figures of merit, such as the error probability, the Q-factor, the eye-opening and so on. Under very strict hypotheses, there is a sort of equivalence among these figures of merit, but things drastically change when nonlinear effects are present or different modulation formats considered.

This depiction of optical communications as an early science is well reflected by the most known journals and conferences of this area, where technological and experimental aspects usually play a predominant role. On the other hand, this book, namely Optical Communications Theory and Techniques, is intended to be a collection of up-to-date papers dealing with the theoretical aspects of optical communications. All the papers were selected or written by worldwide recognized experts in the field, and were presented at the 2004 Tyrrhenian International Workshop on Digital Communications. According to the program of the workshop, the book is divided into four parts:

Information and Communication Theory for Optical Communications. This first part examines optical systems from a rigorous information theory point of view, addressing questions like "what is the ultimate capacity of a given channel?", or "which is the most efficient modulation format?".

Coding Theory and Techniques. This part is concerned with the theory and techniques of coding, applied to optical systems. For instance, different forward error correction (FEC) codes are analyzed and compared, taking explicitly into account the non-AWGN (Additive White Gaussian Noise) nature of the channel.

Characterizing, Measuring, and Calculating Performance in Optical Fiber Communication Systems. This part describes several techniques for the experimental measurement, analytical evaluation or simulations-based estimation of the performance of optical systems. The error probability in the linear and nonlinear regime, as well as the impact of PMD or Raman amplification are subject of this part.

Modulation Formats and Detection. This last part is concerned with the joint or disjoint use of different modulation formats and detection techniques to improve the performance of optical systems and their tolerance to transmission impairments. Modulation in the amplitude, phase and polarization domain are considered, as well as adaptive equalization and maximum likelihood sequence estimation.

Each paper is self contained, such to give the reader a clear picture of the treated topic. Furthermore, getting back to the depiction of optical communications as an early science, the whole book is intended to be a common basis for the theoreticians working in the field, upon which consistent new works could be developed in the next future.

ENRICO FORESTIERI

Acknowledgments

The editor and general chair of the 2004 edition of the Tyrrhenian International Workshop on Digital Communications, held in Pisa on October 2004 as a topical meeting on "Optical Communication Theory and Techniques", is much indebted and wish to express his sincere thanks to the organizers of the technical sessions, namely *Joseph M. Kahn* from Stanford University, USA, *Sergio Benedetto* from Politecnico di Torino, Italy, *Curtis R. Menyuk* from University of Maryland Baltimore County, USA, and *Klaus Petermann* from Technische Universität Berlin, Germany, whose precious cooperation was essential to the organization of the Workshop.

He would also like to thank all the authors for contributing to the Workshop with their high quality papers. Special thanks go to Giancarlo Prati, CNIT director, and to Marco Secondini and Karin Ennser, who generously helped in the preparation of this book.

The Workshop would not have been possible without the support of the Italian National Consortium for Telecommunications (CNIT), and without the sponsorship of the following companies, which are gratefully acknowledged.

I

INFORMATION AND COMMUNICATION THEORY FOR OPTICAL COMMUNICATIONS

SOLVING THE NONLINEAR SCHRÖDINGER EQUATION

Enrico Forestieri and Marco Secondini
Scuola Superiore Sant'Anna di Studi Universitari e Perfezionamento, Pisa, Italy, and Photonic Networks National Laboratory, CNIT, Pisa, Italy.
forestieri@sssup.it

Abstract: Some simple recursive methods are described for constructing asymptotically exact solutions of the nonlinear Schrödinger equation (NLSE). It is shown that the NLSE solution can be expressed analytically by two recurrence relations corresponding to two different perturbation methods.

Key words: optical Kerr effect; optical fiber nonlinearity; nonlinear distortion; optical fiber theory.

1.   INTRODUCTION

The nonlinear Schrödinger equation governs the propagation of the optical field complex envelope $v(z,t)$ in a single-mode fiber [1]. Accounting for group velocity dispersion (GVD), self-phase modulation (SPM), and loss, in a time frame moving with the signal group velocity, the NLSE can be written as

$$\frac{\partial v}{\partial z} = j\frac{\beta_2}{2}\frac{\partial^2 v}{\partial t^2} - j\gamma|v|^2 v - \frac{\alpha}{2}v, \qquad (1)$$

where γ is the Kerr nonlinear coefficient [1], α is the power attenuation constant, and β_2 is the GVD parameter ($\beta_2 = -\lambda^2 D/(2\pi c)$, λ being the reference wavelength, c the light speed, and D the fiber dispersion parameter at λ). Letting $v(z,t) \triangleq e^{-\alpha z/2}u(z,t)$, we can get rid of the last term in (1), which becomes

$$\frac{\partial u}{\partial z} = j\frac{\beta_2}{2}\frac{\partial^2 u}{\partial t^2} - j\gamma e^{-\alpha z}|u|^2 u. \qquad (2)$$

Exact solutions of this equation are typically not known in analytical form, except for soliton solutions when $\alpha = 0$ [2–4]. Given an input condition

$u(0, t)$, the solution of (2) is then to be found numerically, the most widely used method being the *Split-Step Fourier Method* (SSFM) [1]. Analytical approximations to the solution of (2) can be obtained by linearization techniques [5–12], such as perturbation methods taylored for modulation instability (or parametric gain) [5–8] or of more general validity [9, 10], small-signal analysis [11], and the variational method [12]. An approach based on Volterra series [13] was recently shown to be equivalent to the regular perturbation method [9]. However, all methods able to deal with an arbitrarily modulated input signal, provide accurate approximations either only for very small input powers or only for very small fiber losses, with the exception of the *enhanced regular perturbation* method presented in [9] and the multiplicative approximation introduced in [10], whose results are valid for input powers as high as about 10 dBm. We present here two recursive expressions that, starting from the linear solution of (2) for $\gamma = 0$, asymptotically converge to the exact solution for $\gamma \neq 0$, and revisit the multiplicative approximation in [10], relating it to the regular perturbation method.

2.   AN INTEGRAL EXPRESSION OF THE NLSE

In this Section we will obtain an integral expression of the NLSE which, to our knowledge, is not found in the literature. Letting

$$f(z, t) \triangleq e^{-\alpha z}|u(z, t)|^2 u(z, t) \tag{3}$$

and taking the Fourier transform[1] of (2), we obtain

$$\frac{\partial U}{\partial z} = -j\frac{\beta_2}{2}\omega^2 U - j\gamma F, \tag{4}$$

which, by the position

$$U(z, \omega) \triangleq e^{-j\beta_2\omega^2 z/2} Y(z, \omega), \tag{5}$$

becomes

$$\frac{\partial Y}{\partial z} = -j\gamma e^{j\beta_2\omega^2 z/2} F. \tag{6}$$

Integrating (6) from 0 to z leads to

$$Y(z, \omega) = Y(0, \omega) - j\gamma \int_0^z e^{j\beta_2\omega^2\zeta/2} F(\zeta, \omega)d\zeta, \tag{7}$$

and, taking into account (5), we have

$$U(z, \omega) = U_0(z, \omega) - j\gamma \int_0^z e^{-j\beta_2\omega^2(z-\zeta)/2} F(\zeta, \omega)d\zeta, \tag{8}$$

---

[1]The Fourier transform with respect to time t of a function $x(z, t)$ will be denoted by the same but capital letter $X(z, \omega)$, such that $X(z, \omega) = \mathcal{F}\{x(z, t)\}$, and $x(z, t) = \mathcal{F}^{-1}\{X(z, \omega)\}$.

where $U_0(z,\omega) = U(0,\omega)e^{-j\beta_2\omega^2 z/2}$ is the Fourier transform of the solution of (2) for $\gamma = 0$. Letting now $H(z,\omega) \triangleq \exp(-j\beta_2\omega^2 z/2)$, so that $h(z,t) = \mathcal{F}^{-1}\{H(z,\omega)\}$, and antitransforming (8) by taking into account (3), gives

$$u(z,t) = u_0(z,t) - j\gamma \int_0^z \left[|u(\zeta,t)|^2 u(\zeta,t) \right] \otimes h(z-\zeta,t)e^{-\alpha\zeta}d\zeta \qquad (9)$$

where $\otimes$ denotes temporal convolution, and $u_0(z,t) = u(0,t) \otimes h(z,t)$ is the signal at z in a linear and lossless fiber.

3. A FIRST RECURRENCE RELATION CORRESPONDING TO A REGULAR PERTURBATION METHOD

According to the regular perturbation (RP) method [9], expanding the optical field complex envelope $u(z,t)$ in power series in γ

$$u(z,t) = \sum_{k=0}^{\infty} \gamma^k u_k(z,t) \qquad (10)$$

and substituting (10) in (9), after some algebra we obtain

$$\sum_{k=1}^{\infty} \gamma^k u_k = \sum_{n=0}^{\infty} \gamma^{n+1} \left[-j \int_0^z \left(\sum_{k=0}^{n} \sum_{i=0}^{k} u_i u_{k-i} u_{n-k}^* \right) \otimes h(z-\zeta,t)e^{-\alpha\zeta}d\zeta \right] \qquad (11)$$

where we omitted the arguments (z,t) for the u_k's appearing on the left side, and (ζ,t) for those on the right side. By equating the powers in γ with the same exponent, we can recursively evaluate all the u_k's

$$u_n = -j \int_0^z \left(\sum_{k=0}^{n-1} \sum_{i=0}^{k} u_i u_{k-i} u_{n-k}^* \right) \otimes h(z-\zeta,t)e^{-\alpha\zeta}d\zeta, \qquad n \geq 1. \qquad (12)$$

As an example, the first three u_k's turn out to be

$$u_1 = -j \int_0^z \left(|u_0|^2 u_0 \right) \otimes h(z-\zeta,t)e^{-\alpha\zeta}d\zeta,$$

$$u_2 = -j \int_0^z \left(2|u_0|^2 u_1 + u_0^2 u_1^* \right) \otimes h(z-\zeta,t)e^{-\alpha\zeta}d\zeta,$$

$$u_3 = -j \int_0^z \left(2|u_0|^2 u_2 + u_0^2 u_2^* + 2|u_1|^2 u_0 + u_1^2 u_0^* \right) \otimes h(z-\zeta,t)e^{-\alpha\zeta}d\zeta.$$

Turning again our attention to (9), we note that it suggests the following recurrence relation

$$v_0(z,t) = u_0(z,t) \tag{13}$$

$$v_{n+1}(z,t) = u_0(z,t) - j\gamma \int_0^z \left[|v_n(\zeta,t)|^2 v_n(\zeta,t)\right] \otimes h(z-\zeta,t)e^{-\alpha\zeta}d\zeta$$

and it is easy to see that

$$\lim_{n\to\infty} v_n(z,t) = u(z,t) \tag{14}$$

as it can be shown that

$$\frac{1}{k!} \frac{\partial^k v_n(z,t)}{\partial \gamma^k}\bigg|_{\gamma=0} = u_k(z,t), \qquad 0 \le k \le n. \tag{15}$$

This means that the rate of convergence of (13) is not greater than that of (10) when using the same number of terms as the recurrence steps, i.e., it is poor [9]. We will now seek an improved recurrence relation with an accelerated rate of convergence to the solution of (2).

4. AN IMPROVED RECURRENCE RELATION CORRESPONDING TO A LOGARITHMIC PERTURBATION METHOD

As shown in [10], a faster convergence rate is obtained when expanding in power series in γ the log of $u(z,t)$ rather than $u(z,t)$ itself as done in (10). So, we try to recast (9) in terms of $\log u(z,t)$, and to this end rewrite it as

$$\frac{u(z,t) - u_0(z,t)}{u_0(z,t)} = -\frac{j\gamma}{u_0(z,t)} \int_0^z \left[|u(\zeta,t)|^2 u(\zeta,t)\right] \otimes h(z-\zeta,t)e^{-\alpha\zeta}d\zeta. \tag{16}$$

Using now the expansion

$$\log \frac{u}{u_0} = \frac{u-u_0}{u_0} - \frac{1}{2}\left(\frac{u-u_0}{u_0}\right)^2 + \frac{1}{3}\left(\frac{u-u_0}{u_0}\right)^3 - \dots \tag{17}$$

we replace the term $(u - u_0)/u_0$ in (16), obtaining

$$\log \frac{u}{u_0} = -\frac{j\gamma}{u_0} \int_0^z [|u|^2 u] \otimes h\, e^{-\alpha\zeta}d\zeta - \frac{1}{2}\left(\frac{u-u_0}{u_0}\right)^2 + \frac{1}{3}\left(\frac{u-u_0}{u_0}\right)^3 - \dots \tag{18}$$

where, for simplicity, we omitted all the function arguments. So, the sought improved recurrence relation suggested by (18) is

$$v_0 = u_0 \tag{19}$$

$$v_{n+1} = v_n \exp\left\{-\frac{j\gamma}{v_0} \int_0^z [|v_n|^2 v_n] \otimes h(z-\zeta,t)e^{-\alpha\zeta}d\zeta - \frac{v_n - v_0}{v_0}\right\}$$

where we used again (17) to obtain the right side of the second equation. Also in this case $\lim_{n\to\infty} v_n(z,t) = u(z,t)$, as it can be shown that the power series in γ of $\log v_n(z,t)$ coincides with that of $\log u(z,t)$ in the first n terms.

Notice that $v_1(z,t)$ evaluated from (19) coincides with the first-order multiplicative approximation in [10], there obtained with a different approach. The method in [10] is really a logarithmic perturbation (LP) method as the solution $u(z,t)$ is written as

$$u(z,t) = u_0(z,t)e^{-j\gamma\vartheta(z,t)}, \quad \vartheta = \vartheta_0 + \gamma\vartheta_1 + \gamma^2\vartheta_2 + \dots \quad (20)$$

and $\vartheta_0(z,t)$, $\vartheta_1(z,t)$ are evaluated by analytically approximating the SSFM solution. The calculation of $\vartheta_n(z,t)$ becomes progressively more involved for increasing values of n, but that method is useful because it can provide an analytical expression for the SSFM errors due to a finite step size [10].

We now follow another approach. Letting

$$\psi_n(z,t) \stackrel{\triangle}{=} -j\vartheta_{n-1}(z,t), \quad (21)$$

such that

$$u(z,t) = u_0(z,t)\exp\left(\gamma\psi_1(z,t) + \gamma^2\psi_2(z,t) + \gamma^3\psi_3(z,t) + \dots\right), \quad (22)$$

for every n, $\psi_n(z,t)$ can be easily evaluated in the following manner. The power series expansion of $u(z,t)$ in (22) is

$$u_0\exp\left(\sum_{k=1}^{\infty}\gamma^k\psi_k\right) = u_0\sum_{i=0}^{\infty}\frac{1}{i!}\left(\sum_{k=1}^{\infty}\gamma^k\psi_k\right)^i$$

$$= u_0\left[1 + \sum_{n=1}^{\infty}\left(\sum_{k=1}^{n}\frac{\varphi_{n,k}}{k!}\right)\gamma^n\right] \quad (23)$$

where

$$\varphi_{n,k} = \begin{cases} 0 & \text{if } n \neq 0, k = 0 \\ 1 & \text{if } n = k = 0 \\ \displaystyle\sum_{m=k-1}^{n-1}\varphi_{m,k-1}\psi_{n-m} & \text{otherwise} \end{cases} \quad (24)$$

Equating (10) to (23), and taking into account that $\varphi_{n,1} = \psi_n$, we can recursively evaluate the ψ_n's as

$$\psi_n = \frac{u_n}{u_0} - \sum_{k=2}^{n}\frac{\varphi_{n,k}}{k!}. \quad (25)$$

Thus, from the n-th order regular perturbation approximation we can construct the n-th order logarithmic perturbation approximation. As an example we have

$$\psi_1 = \frac{u_1}{u_0},$$

$$\psi_2 = \frac{u_2}{u_0} - \frac{1}{2}\psi_1^2,$$

$$\psi_3 = \frac{u_3}{u_0} - \psi_1\psi_2 - \frac{1}{6}\psi_1^3,$$

$$\psi_4 = \frac{u_4}{u_0} - \psi_1\psi_3 - \frac{1}{2}\psi_2^2 - \frac{1}{2}\psi_1^2\psi_2 - \frac{1}{24}\psi_1^4.$$

So, once evaluated the u_k's from (12), we can evaluate the ψ_k's from (25) and then $u(z,t)$ through (22), unless $u_0(z,t)$ is zero (or very small), in which case we simply use (10) as in this case $u(z,t)$ is also small and (10) is equally accurate.

5.   COMPUTATIONAL ISSUES

The computational complexity of (12), (13) and (19) is the same, and at first glance it may seem that a n-th order integral must be computed for the n-th order approximation. However, it is not so and the complexity only increases linearly with n. Indeed, the terms depending on z can be taken out of the integration[2] and so all the integrals can be computed in parallel. However, only for $n \leq 2$ these methods turn out to be faster than the SSFM because of the possibility to exploit efficient quadrature rules for the outer integral, whereas the inner ones are to be evaluated through the trapezoidal rule as, to evaluate them in parallel, we are forced to use the nodes imposed by the outer quadrature rule.

Although (12), (13), (19), and (22) hold for a single fiber span, they can also be used in the case of many spans with given dispersion maps and per-span amplification. Indeed, one simply considers the output signal at the end of each span as the input signal to the next span [9, 10]. We would like to point out that even if the propagation in the compensating fiber is considered to be linear, (19) or (22) should still be used for the total span length L, by simply replacing z with the length of the transmission fiber L_F in the upper limit of integration and with L in all other places.

---

[2]This is apparent when performing the integrals in the frequency domain, but is also true in the time domain as $h(z - \zeta, t) = h(z,t) \otimes h(-\zeta, t)$ when $h(\ell, t)$ is the impulse response of a linear fiber of length ℓ ($h(-\zeta, t)$ simply corresponds to a fiber of length ζ and opposite sign of dispersion parameter).

6.   SOME RESULTS

To illustrate the results obtainable by the RP and LP methods, we considered a $n \times 100$ km link, composed of n 100 km spans of transmission fiber followed by a compensating fiber and per span amplification recovering all the span loss. The transmission fiber is a standard single-mode fiber with $\alpha = 0.19$ dB/km, $D = 17$ ps/nm/km, $\gamma = 1.3$ W^{-1}km^{-1}, whereas the compensating fiber has $\alpha = 0.6$ dB/km, $D = -100$ ps/nm/km, $\gamma = 5.5$ W^{-1}km^{-1}, and a length such that the residual dispersion per span is zero.

In Table 1 we report the minimum order of the RP and LP methods necessary to have a normalized square deviation (NSD) less than 10^{-3}. The NSD is defined as

$$\text{NSD} = \frac{\int |u_{SSFM}(z,t) - u_P(z,t)|^2 \, dt}{\int |u_{SSFM}(z,t)|^2 \, dt} \tag{26}$$

where $u_{SSFM}(z,t)$ is the solution obtained by the SSFM with a step size of 100 m, $u_P(z,t)$ is either the RP or LP approximation, and the integrals extend to the whole transmission period, which in our case is that corresponding to a pseudorandom bit sequence of length 64 bits. The input signal format is NRZ at 10 Gb/s, filtered by a Gaussian filter with bandwidth equal to 20 GHz.

Table 1.   Minimum order of the RP and LP methods necessary to achieve NSD $< 10^{-3}$ for a given input peak power and number of spans.

Spans	3 dBm		6 dBm		9 dBm		12 dBm	
	RP	LP	RP	LP	RP	LP	RP	LP
1	1	1	1	1	1	1	2	1
2	1	1	1	1	2	1	3	2
3	1	1	1	1	2	1	3	2
4	1	1	1	1	2	1	3	2
5	1	1	2	1	2	1	3	2
6	1	1	2	1	2	1	3	2
7	1	1	2	1	2	2	3	2
8	1	1	2	1	3	2	3	2
9	1	1	2	1	3	2	3	3
10	1	1	2	1	3	2	4	4

It can be seen that the LP method requires a lower order than the RP method to achieve the same accuracy when the input peak power P_{in} increases beyond 6 dBm and the number of spans execeeds 4. As an example, Fig. 1 shows the output intensity for an isolated "1" in the pseudorandom sequence when the input peak power is 12 dBm and the number of spans is 5, showing that, in this case, 3rd-order is required for the RP method, whereas only 2nd-order for the LP method. As a matter of fact, until 12 dBm and 8 spans, the 2nd-order LP method suffices for a NSD $< 10^{-3}$. However, for higher values of P_{in}

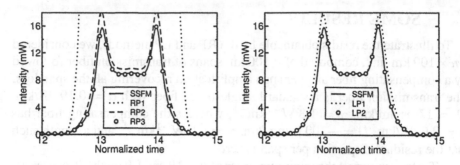

Figure 1.   Output intensity for an isolated "1" with $P_{in} = 12$ dBm and 5 spans.

and number of spans, i.e., when moving form left to right along a diagonal in Table 1, the two methods tend to become equivalent, in the sense that they tend to require the same order to achieve a given accuracy.

This can be explained by making the analytical form (19) of the NLSE solution explicit. Indeed, doing so we can see that the nonlinear parameter γ appears at the exponent, and then at the exponent of the exponent, and then at the exponent of the exponent of the exponent, and so on. So, the LP approximation has an initial advantage over the RP one, but when orders higher than 3 or 4 are needed, this initial advantage is lost and the two approximations tend to coincide.

7.   CONCLUSIONS

We presented two recurrence relations that asymptotically approach the solution of the NLSE. Although they represent an analytical expression of such solution, their computational complexity increases linearly with the recursion depth, making them not practical for a too high order of recursion. Nevertheless, for practical values of input power and number of spans, as those used in current dispersion managed systems, the second-order LP method can provide accurate results in a shorter time than the SSFM (we estimated an advantage of about 30% for approximately the same accuracy). Furthermore, we believe that these expressions can have a theoretical value, for example in explaining that the RP and LP methods are asymptotically equivalent, as we did.

REFERENCES

[1] *Nonlinear fiber optics*. San Diego: Academic Press, 1989.

[2] V. E. Zakharov and A. B. Shabat, "Exact theory for two-dimensional self-focusing and one-dimensional self-modulation of waves in nonlinear media," *Sov. Phys. JETP*, vol. 34, pp. 62–69, 1972.

[3] N. N. Akhmediev, V. M. Elonskii, and N. E. Kulagin, "Generation of periodic trains of picosecond pulses in an optical fiber: exact solution," *Sov. Phys. JETP*, vol. 62, pp. 894–

899, 1985.

[4] E. R. Tracy, H. H. Chen, and Y. C. Lee, "Study of quasiperiodic solutions of the nonlinear schrodinger equation and the nonlinear modulational instability," *Phys. Rev. Lett.*, vol. 53, pp. 218–221, 1984.

[5] M. Karlsson, "Modulational instability in lossy optical fibers," *J. Opt. Soc. Am. B*, vol. 12, pp. 2071–2077, Nov. 1995.

[6] A. Carena, V. Curri, R. Gaudino, P. Poggiolini, and S. Benedetto, "New analytical results on fiber parametric gain and its effects on ASE noise," *IEEE Photon. Technol. Lett.*, vol. 9, pp. 535–537, Apr. 1997.

[7] R. Hui, M. O'Sullivan, A. Robinson, and M. Taylor, "Modulation instability and its impact in multispan optical amplified imdd systems: theory and experiments," *J. Lightwave Technol.*, vol. 15, pp. 1071–1082, July 1997.

[8] C. Lorattanasane and K. Kikuchi, "Parametric instability of optical amplifier noise in long-distance optical transmission systems," *IEEE J. Quantum Electron.*, vol. 33, pp. 1058–1074, July 1997.

[9] A. Vannucci, P. Serena, and A. Bononi, "The RP method: a new tool for the iterative solution of the nonlinear Schrödinger equation," *J. Lightwave Technol.*, vol. 20, pp. 1102–1112, July 2002.

[10] E. Ciaramella and E. Forestieri, "Analytical approximation of nonlinear distortions," *IEEE Photon. Technol. Lett.*, 2004. To appear.

[11] A. V. T. Cartaxo, "Small-signal analysis for nonlinear and dispersive optical fibres, and its application to design of dispersion supported transmission systems with optical dispersion compensation," *IEE Proc.-Optoelectron.*, vol. 146, pp. 213–222, Oct. 1999.

[12] H. Hasegawa and Y. Kodama, *Solitons in Optical Communications*. New York: Oxford University Press, 1995.

[13] K. V. Peddanarappagari and M. Brandt-Pearce, "Volterra series transfer function of single-mode fibers," *IEEE J. Lightwave Technol.*, vol. 15, pp. 2232–2241, Dec. 1997.

MODULATION AND DETECTION TECHNIQUES FOR DWDM SYSTEMS*

Invited Paper

Joseph M. Kahn[1] and Keang-Po Ho[2]

[1]*Stanford University, Department of Electrical Engineering, Stanford, CA 94305 USA, e-mail: jmk@ee.stanford.edu;* [2]*National Taiwan University, Institute of Communication Engineering and Department of Electrical Engineering, Taipei 106, Taiwan, e-mail: kpho@cc.ee.ntu.edu.tw*

Abstract: Various binary and non-binary modulation techniques, in conjunction with appropriate detection techniques, are compared in terms of their spectral efficiencies and signal-to-noise ratio requirements, assuming amplified spontaneous emission is the dominant noise source. These include (a) pulse-amplitude modulation with direct detection, (b) differential phase-shift keying with interferometric detection, (c) phase-shift keying with coherent detection, and (d) quadrature-amplitude modulation with coherent detection.

Key words: optical fiber communication; optical modulation; optical signal detection; differential phase-shift keying; phase-shift keying; pulse amplitude modulation; heterodyning; homodyne detection.

1. INTRODUCTION

Currently deployed dense wavelength-division-multiplexed (DWDM) systems use binary on-off keying (OOK) with direct detection. In an effort to improve spectral efficiency and robustness against transmission impairments, researchers have investigated a variety of binary and non-binary modulation techniques, in conjunction with various detection techniques. In this paper, we compare the spectral efficiencies and signal-to-noise ratio (SNR) requirements of several modulation and detection techniques. We assume that amplified spontaneous emission (ASE) from optical amplifiers is the dominant noise

*This research was supported at Stanford University by National Science Foundation Grant ECS-0335013 and at National Taiwan University by National Science Council of R.O.C. Grant NSC-92-2218-E-002-034.

source. We do not explicitly consider the impact of other impairments, such as fiber nonlinearity (FNL), chromatic dispersion (CD), or polarization-mode dispersion (PMD).

The information bit rate per channel in one polarization is given by

$$R_b = R_s R_c \log_2 M , \tag{1}$$

where R_s is the symbol rate, $R_c \leq 1$ is the rate of an error-correction encoder used to improve SNR efficiency, and M is the number of transmitted signals that can be distinguished by the receiver. For an occupied bandwidth per channel B, avoidance of intersymbol interference requires $R_s \leq B$ [1]. If the channel spacing is Δf, the spectral efficiency per polarization is

$$S = \frac{R_b}{\Delta f} = \frac{R_s R_c \log_2 M}{\Delta f} \leq \frac{B R_c \log_2 M}{\Delta f} \tag{2}$$

Our figure of merit for spectral efficiency is $\log_2 M$, the number of coded bits per symbol, which determines spectral efficiency at fixed $R_s/\Delta f$ and fixed R_c. Binary modulation ($M = 2$) can achieve spectral efficiency up to 1 b/s/Hz, while non-binary modulation ($M > 2$) can achieve higher spectral efficiencies.

Non-binary modulation can improve tolerance to uncompensated CD and PMD, as compared to binary modulation, for two reasons [2, 3]. At a given bit rate R_b, non-binary modulation can employ lower symbol rate R_s, reducing signal bandwidth B, thus reducing pulse spreading caused by CD. Also, because non-binary modulation employs longer symbol interval $1/R_s$, it can often tolerate greater pulse spreading caused by CD and PMD.

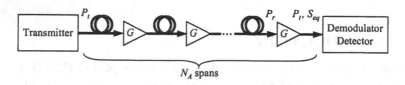

Figure 1.   Equivalent block diagram of multi-span system.

In comparing SNR efficiencies, we consider the reference system shown in Fig. 1. The system comprises N_A fiber spans, each of gain $1/G$, and each followed by an amplifier of gain G. The average transmitted power per channel is P_t, while the average power at the input of each amplifier is $P_r = P_t/G$. We assume that for all detection schemes, ASE dominates over other noise sources, thereby maximizing the receiver signal-to-noise ratio (SNR) [4]. At the output of the final amplifier, the ASE in one polarization has a power spectral density

(PSD) given by

$$S_{eq} = N_A(G - 1)n_{sp}h\nu = (G - 1)n_{eq}h\nu, \qquad (3)$$

where n_{sp} is the spontaneous emission noise factor of one amplifier, and we define the equivalent noise factor of the multi-span system by $n_{eq} = N_A n_{sp}$.

At the input of the final amplifier, the average energy per information bit is $E_b = P_r/R_b$. At the output of the final amplifier, the average energy per information bit is $GE_b = GP_r/R_b = P_t/R_b$, identical to the average transmitted energy per information bit. Our figure of merit for SNR efficiency is the value of the received SNR per information bit GE_b/S_{eq} required to achieve an information bit-error ratio (BER) $P_b = 10^{-9}$. This figure of merit indicates the average energy that must be transmitted per information bit for fixed ASE noise, making it appropriate for systems in which transmitted energy is constrained by FNL. Defining the average number of photons per information bit at the input of the final amplifier $n_b = E_b/h\nu$, and using (3), the figure of merit for SNR efficiency is

$$\frac{GE_b}{S_{eq}} = \left(\frac{G}{G-1}\right)\frac{n_b}{n_{eq}} \approx \frac{n_b}{n_{eq}}, \qquad (4)$$

which is equal to the receiver sensitivity at the final amplifier input divided by the equivalent noise factor of the multi-span system.

The modulation techniques described below can be employed with various elementary pulse shapes, including non-return-to-zero (NRZ) or return-to-zero (RZ), and with various line codes, such as duobinary or carrier-suppressed RZ. In the absence of fiber nonlinearity, with proper CD compensation and matched filtering, the elementary pulse shape and line code do not affect the spectral efficiency and SNR figures of merit considered here.

2.   DIRECT DETECTION OF PAM

When used with direct detection, M-ary pulse-amplitude modulation (PAM) encodes a block of $\log_2 M$ bits by transmitting one of M intensity levels. Henry [5] and Humblet and Azizoglu [6] analyzed the performance of 2-PAM (OOK) with optical preamplification and direct detection. In order to achieve $P_b = 10^{-9}$, 2-PAM requires $n_b/n_{eq} = 38$ with single-polarization filtering and $n_b/n_{eq} = 41$ with polarization diversity.

We are not aware of an exact performance analysis of M-PAM for $M \geq 4$. Neglecting all noises except the dominant signal-spontaneous beat noise, at each intensity level, the photocurrent is Gaussian-distributed, with a variance proportional to the intensity. Setting the $M - 1$ decision thresholds at the geometric means of pairs of adjacent levels approximately equalizes the downward and upward error probabilities at each threshold. In order to equalize the error probabilities at the $M - 1$ different thresholds, the M intensity levels should form a quadratic series [7]. Assuming Gray coding, the BER is given approximately by

$$P_b \approx \frac{1}{\log_2 M} Q \left(\sqrt{\frac{3 \log_2 M}{(2M - 1)(M - 1)} \frac{GE_b}{S_{eq}}} \right)$$

$$= \frac{1}{\log_2 M} Q \left(\sqrt{\frac{3 \log_2 M}{(2M - 1)(M - 1)} \frac{n_b}{n_{eq}}} \right). \tag{5}$$

For $M = 2$, (5) indicates that $n_b/n_{eq} = 36$ is required for $P_b = 10^{-9}$, which is lower by 0.2 dB than the exact requirement $n_b/n_{eq} = 38$. For $M \geq 4$, (5) indicates that the SNR requirement increases by a factor $(3 \log_2 M)/[(2M - 1)(M - 1)]$, corresponding to penalties of 5.5, 10.7 and 15.9 dB for $M = 4$, 8, 16, respectively. To estimate SNR requirements of M-PAM with single-polarization filtering, we assume the exact requirement $n_b/n_{eq} = 38$ for $M = 2$, and add the respective penalties for $M = 4, 8, 16$.

3.   INTERFEROMETRIC DETECTION OF DPSK

Both M-ary phase-shift keying (PSK) and differential phase-shift keying (DPSK) use signal constellations consisting of M points equally spaced on a circle. While M-PSK encodes each block of $\log_2 M$ bits in the phase of the transmitted symbol, M-DPSK encodes each block of $\log_2 M$ bits in the phase change between successively transmitted symbols [1].

For interferometric detection of 2-DPSK, a Mach-Zehnder interferometer with a delay difference of one symbol compares the phases transmitted in successive symbols, yielding an intensity-modulated output that is detected by a balanced optical receiver. In the case of M-DPSK, $M \geq 4$, a pair of Mach-Zehnder interferometers (with excess phase shifts of 0 and $\pi/2$) and a pair of balanced receivers are used to determine the in-phase and quadrature components of the phase change between successive symbols.

Tonguz and Wagner [8] showed that the performance of DPSK with optical amplification and interferometric detection is equivalent to standard differentially coherent detection [1]. 2-DPSK requires $n_b/n_{eq} = 20$ with single-polarization filtering and $n_b/n_{eq} = 22$ with polarization diversity to achieve $P_b = 10^{-9}$ [8]. The performance of M-DPSK for $M \geq 4$ with single-polarization polarization filtering is described by the analysis in [1].

4.   COHERENT DETECTION OF PSK AND QAM

In optical communications, "coherent detection" has often been used to denote any detection process involving photoelectric mixing of a signal and a local oscillator [9]. Historically, the main advantages of coherent detection were considered to be high receiver sensitivity and the ability to perform channel demultiplexing and CD compensation in the electrical domain [9]. From a current perspective, the principal advantage of coherent detection is the ability to detect information encoded independently in both in-phase and quadrature field components, increasing spectral efficiency. This advantage can be achieved only by using synchronous detection, which requires an optical or electrical phase-locked loop (PLL), or some other carrier-recovery technique. Hence, we use the term "coherent detection" only to denote synchronous detection, which is consistent with its use in non-optical communications [1].[1]

In ASE-limited systems, the sensitivity of a synchronous heterodyne receiver is equivalent to a synchronous homodyne receiver provided that the ASE is narrow-band-filtered or that image rejection is employed [10]. Most DWDM systems use demultiplexers that provide narrow-band filtering of the received signal and ASE, in which case, image rejection is not required for heterodyne to achieve the same performance as homodyne detection.

Both homodyne and heterodyne detection require polarization tracking or polarization diversity. Our analysis assumes tracking, as it requires fewer photodetectors. Coherent system performance is optimized by using high amplifier gain G and a strong local-oscillator laser, so that local-oscillator-ASE beat noise dominates over receiver thermal noise and other noise sources [4]. This corresponds to the standard case of additive white Gaussian noise [1].

M-ary PSK uses a constellation consisting of M points equally spaced on a circle. In the case of uncoded 2- or 4-PSK, the BER is given by [1]

$$P_b = Q\left(\sqrt{\frac{2GE_b}{S_{eq}}}\right) = Q\left(\sqrt{\frac{2n_b}{n_{eq}}}\right), \tag{6}$$

where the Q function is defined in [1]. Achieving a BER 10^{-9} requires $n_b/n_{eq} = 18$. The BER performance of M-PSK, $M > 4$ is computed in [1].

M-ary quadrature-amplitude modulation (QAM) uses a set of constellation points that are roughly uniformly distributed within a two-dimensional region. In the cases $M = 2^{2m}$ ($M = 4, 16, \ldots$), the points are evenly arrayed in a

---

[1] We do not consider heterodyne or phase-diversity homodyne detection with differentially coherent (delay) demodulation of DPSK, since the interferometric detection scheme described in Section 3 is mathematically equivalent [8] and is easier to implement. Likewise, we do not consider heterodyne or phase-diversity homodyne detection with noncoherent (envelope) demodulation of PAM, since the direct detection scheme described in Section 2 is mathematically equivalent [8] and is more easily implemented.

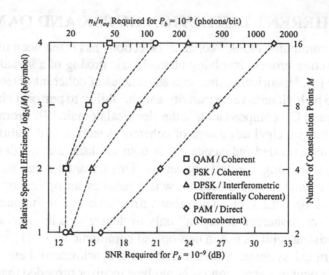

Figure 2.   Spectral efficiency vs. SNR requirement for various techniques.

Table 1.   Comparison of modulation and detection schemes. Numbers given represent the values of $GE_b/S_{eq} = n_b/n_{eq}$ (photons/bit) required for $P_b = 10^{-9}$. Numbers in parenthesis are the corresponding values of $10 \log_{10}(GE_b/S_{eq}) = 10 \log_{10}(n_b/n_{eq})$.

M	$\log_2 M$	PSK/Coherent One Pol.	QAM/Coherent One Pol.	DPSK/Interferometric		PAM/Direct	
				One Pol.	Two Pol.	One Pol.	Two Pol.
2	1	18 (12.6)	Not applicable	20 (13.0)	22 (13.4)	38 (15.8)	41 (16.1)
4	2	18 (12.6)	18 (12.6)	31 (14.9)	?	134 (21.3)	?
8	3	41 (16.2)	29 (14.6)	83 (19.2)	?	443 (26.5)	?
16	4	119 (20.8)	45 (16.6)	240 (23.8)	?	1472 (31.7)	?

$2^m \times 2^m$ square, while in the cases $M = 2^{2m+1}$ ($M = 8, 32, \ldots$), the points are often arranged in a cross. The BER performance of M-QAM is computed in [1].

5.   DISCUSSION

Fig. 2 and Table 1 compare the spectral efficiencies and SNR requirements of the various modulation and detection techniques described above. We observe that for $M > 2$, the SNR requirement for PAM increases very rapidly, while the SNR requirements of the other three techniques increase at a more moderate rate. Note that for large M, the SNR requirements increase with roughly equal slopes for PAM, DPSK and PSK, while QAM exhibits a distinctly slower increase of SNR requirement. This behavior can be traced to

Table 2. Comparison of detection techniques. Shading denotes an advantage.

Attribute	Direct	Interferometric	Coherent
Maximum degrees of freedom per polarization	1	1	2
Signal-to-noise requirement for binary modulation (relative to 2-PAM with direct detection)	0 dB (2-PAM)	−2.8 dB (2-DPSK)	−3.2 dB (2-PSK)
Signal-to-noise requirement for quaternary modulation (relative to 2-PAM with direct detection)	+5.5 dB (4-PAM)	−0.9 dB (4-DPSK)	−3.2 dB (4-PSK)
Electrical filtering can select WDM channel	No	No	Yes
Chromatic dispersion is linear distortion, making electrical compensation more effective	No	No	Yes
Local oscillator laser required at receiver	No	No	Yes
Polarization control or diversity required	No	No	Yes

the fact that PAM, DPSK and PSK offer one degree of freedom per polarization (either magnitude or phase), while QAM offers two degrees of freedom per polarization (both in-phase and quadrature field components). Based on Fig. 2, at spectral efficiencies below 1 b/s/Hz per polarization, 2-PAM (OOK) and 2-DPSK are attractive techniques. Between 1 and 2 b/s/Hz, 4-DPSK and 4-PSK are perhaps the most attractive techniques. At spectral efficiencies above 2 b/s/Hz, 8-PSK and 8- and 16-QAM become the most attractive techniques.

Table 2 compares key attributes of direct, interferometric and coherent detection. The key advantages of interferometric detection over direct detection lie in the superior SNR efficiency of 2- and 4-DPSK as compared to 2- and 4-PAM. Coherent detection is unique in offering two degrees of freedom per polarization, leading to outstanding SNR efficiency for 2- and 4-PSK, and still reasonable SNR efficiency for 8-PSK and for 8- and 16-QAM. Coherent detection also enables electrical channel demultiplexing and CD compensation. Coherent detection requires a local oscillator laser and polarization control, which are significant drawbacks.

Laser phase noise has traditionally been a concern for optical systems using DPSK, PSK or QAM. Interferometric detection of DPSK can be impaired by changes in laser phase between successive symbols. In coherent detection of PSK or QAM, a PLL (optical or electrical) attempts to track the laser phase noise, but the PLL operation is corrupted by ASE noise. Linewidth requirements for 2-DPSK, 2-PSK and 4-PSK are summarized in Table 3. At a bit rate $R_b = 10$ Gb/s, the linewidth requirements for 2-DPSK and 2-PSK can be accommodated by standard distributed-feedback lasers. 4-PSK requires a much narrower linewidth, which can be achieved by compact external cavity lasers [14].

Table 3. Laser linewidth requirements for various modulation and detection techniques, assuming a 0.5 dB penalty. For interferometric detection, transmitter has linewidth $\Delta\nu$, while for coherent detection, each of the transmitter and local oscillator has linewidth $\Delta\nu$.

Modulation	Detection	$\Delta\nu/R_b$	$\Delta\nu$ for $R_b = 10$ Gb/s	Reference
2-DPSK	Interferometric	3.0×10^{-3}	30 MHz	[11]
4-DPSK	Interferometric	?	?	
2-PSK	Coherent	8.0×10^{-4}	8 MHz	[12]
4-PSK	Coherent	2.5×10^{-5}	250 kHz	[13]

REFERENCES

[1] J. G. Proakis, *Digital Communications, 4th Ed.*, McGraw-Hill, 2000.

[2] S. Walklin and J. Conradi, "Multilevel signaling for increasing the reach of 10 Gb/s lightwave systems", *J. of Lightwave Technol.*, vol. 17, pp. 2235–2248, 1999.

[3] J. Wang and J. M. Kahn, "Impact of chromatic and polarization-mode dispersions on DPSK systems using interferometric demodulation and direct detection", *J. Lightwave Technol.* vol. 22, no. 2, pp. 362–371, Feb. 2004.

[4] E. Desurvire, *Erbium-Doped Fiber Amplifiers: Principles and Applications*, Wiley, 1994.

[5] P. S. Henry, "Error-rate performance of optical amplifiers", *Proc. of Conf. on Optical Fiber Commun.*, Washington, DC, 1989, p. 170.

[6] P. A. Humblet and M. Azizoglu, "On the bit error rate of lightwave systems with optical amplifiers", *J. Lightwave Technol.*, vol. 9, pp. 1576–1582, 1991.

[7] J. Rebola and A. Cartaxo, "Optimization of level spacing in quaternary optical communication systems", *Proc. of SPIE*, vol. 4087, pp. 49–59, 2000.

[8] O. K. Tonguz and R. E. Wagner, "Equivalence between preamplified direct detection and heterodyne receivers", *IEEE Photon. Technol. Lett.*, vol. 3, pp. 835-837, 1991.

[9] G. P. Agrawal, *Fiber Optic Communication Systems, 3rd Ed.*, Wiley, 2002.

[10] B. F. Jorgensen, B. Mikkelsen and C. J. Mahon, "Analysis of optical amplifier noise in coherent optical communication systems with optical image rejection receivers", *J. of Lightwave Technol.*, vol. 10, pp. 660–671, 1992.

[11] C. P. Kaiser, P. J. Smith and M. Shafi, "An improved optical heterodyne DPSK receiver to combat laser phase noise, *J. Lightwave Technol*, vol. 13, pp. 525–533, Mar. 1995.

[12] S. Norimatsu and K. Iwashita, "Linewidth requirements for optical synchronous detection systems with nonnegligible loop delay time", *J. Lightwave Technol.*, vol. 10, pp. 341–349, Mar. 1992.

[13] J. R. Barry and J. M. Kahn, "Carrier Synchronization for Homodyne and Heterodyne Detection of Optical Quadriphase-Shift Keying", *J. Lightwave Technol.*, vol. 10, pp. 1939–1951, Dec. 1992.

[14] J. D. Berger, Y. Zhang, J. D. Grade, H. Lee, S. Hrinya, H. Jerman, A. Fennema, A. Tselikov, and D. Anthon, "Widely tunable external cavity diode laser using a MEMS electrostatic rotary actuator", *Proc. of 27th Euro. Conf. on Optical Commun.* Amsterdam, Netherlands, Sept. 30-Oct. 4, 2001.

BEST OPTICAL FILTERING FOR DUOBINARY TRANSMISSION
Invited Paper

G. Bosco, A. Carena, V. Curri, and P. Poggiolini
Dipartimento di Elettronica, Politecnico di Torino, C.so Duca degli Abruzzi, 24 - 10129 Torino - Italy. E-mail:[lastname]@polito.it Tel. +39-011-5644036 Fax +39-011-5644099.

Abstract: We show that for optical transmission systems based on duobinary line-coding, in general the optimum receiver is not based on the optical filter matched to transmitted pulse-shape. In general, the receiver optical filter must be optimized for each transmitted pulse within the ISI conditions imposed by the duobinary line-coding. In order to achieve such a result, we have derived the expression of the parameter K to be maximized with the purpose to decide the optimal filter for each pulse-shape.

Key words: optical fiber communication; modulation formats; duobinary coding; quantum limit; optical filters.

1. INTRODUCTION

The duobinary format was first proposed in the 60's for radio communications [1]. Its high spectral efficiency was the aspect that made it attractive in that context. Later, duobinary was overcome by multilevel schemes that could reach an even higher bandwidth efficiency. Duobinary has recently re-emerged in the field of optical communications. Different implementations have been proposed, among which [2–5]. Comprehensive review papers on the advantages and disadvantages of the use of optical duobinary have been published, such as [6]. It has been pointed out that duobinary, besides a high bandwidth efficiency, also features a very high resilience to fiber chromatic dispersion.

Regarding the sensitivity performance of duobinary, diverging opinions exist. In [2] it was shown that a specific receiver performed in back-to-back equally well with either conventional IMDD or duobinary, suggesting a similar performance of the two formats. A more commonly acknowledged notion is

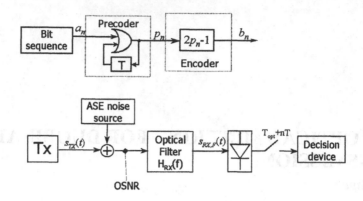

Figure 1.   Duobinary transmitter architecture (top) and analyzed back-to-back system layout (bottom).

that duobinary may have a sensitivity penalty with respect to IMDD. In [7] we presented a rigorous analysis of the ASE-noise-limited, back-to-back sensitivity performance of duobinary, showing that the *quantum limit* [8] of duobinary is at least 0.91 dB *better* than that of IMDD.

After briefly recalling the derivation of such fundamental limit in Section 2, we focus on the pulse dependence of the bit-error rate which is a peculiar characteristic of duobinary transmission. In communication theory it has been shown that the optimum coherent receiver for intensity modulation systems is based on a filter matched to the transmitted pulse [9]. In general, this is valid also for optical systems based on intensity-modulation direct-detection (IMDD), even though a quadratic detector is used to perform optical-to-electrical conversion of the signal [10]. In this case the matched filter is the band-pass optical filter preceding the photodetector. In this work, we consider a simple use-case based on rectangular pulses and filter responses, and demonstrate that when choosing duobinary line-coding the matched-filter assumption is not valid in general. Moreover, we define the parameter K to be maximized in order to obtain the optimal receiver filter for a given pulse shape. For this parameter we report the analytical expression that can be used for any pulse-shape.

2.   QUANTUM LIMIT FOR DUOBINARY

The duobinary TX structure (shown in Fig. 1) that can be found in early papers [11,12] and in textbooks [9] is composed by a precoder, which transforms the information bit sequence a_n into a new bit sequence $p_n = a_n \oplus p_{n-1}$, where the symbol $\oplus$ represents a logical *xor* operation, followed by the processing: $b_n = 2p_n - 1$. The bipolar sequence $b_n \in \{-1, 1\}$ is then used to create the transmitted signal:

$$\mathbf{s}_{TX}(t) = \sqrt{P_{S,av}} \sum_n b_n u\,(t - nT)\; e^{\jmath 2\pi f_0 t}\hat{v}_p. \tag{1}$$

where $P_{S,av}$ is the average power of $\mathbf{s}_{TX}(t)$. $u(t)$ is the normalized transmission pulse (with unitary power), T is the inverse of the bit-rate R_B and $\hat{v}_p$ is the complex unit vector defining the polarization of the modulated signal. This signal can be either received using a coherent receiver or through direct-detection.

In optical communications, duobinary transmission is typically obtained taking advantage of the Mach-Zehnder modulator phase properties and of narrow electric filtering: it is called PSBT [4]. On the receiver side, a standard IMDD receiver is employed. We analyzed an optical duobinary system limited by ASE noise in back-to-back configuration as shown in Fig. 1. In [7], the duobinary application to optical communications has been analyzed showing that the received optical signal $s_{RX}(t)$ after the optical filter at the optimum sampling instant can be written as:

$$\mathbf{s}_{RX}(t_{opt}) = \left\{ \sqrt{P_{S,av}}c_n x(0) + n_{pF}(t) \right\} \hat{v}_p + n_{oF}(t)\,\hat{v}_o. \tag{2}$$

where $c_n = 0$ if $b_n \neq b_{n-1}$ (i.e., $a_n = 0$) and $c_n = \pm 2$ if $b_n = b_{n-1}$ (i.e., $a_n = 1$). $n_{oF}(t)\,\hat{v}_o$ is the noise component on the polarization orthogonal to the modulated signal. Note that the received pulse $x(t)$ must comply with the duobinary ISI condition, i.e., $x(0) = x(T) \neq 0$ and $x(nT) = 0\;\;\forall n \neq 0, 1$.

The received optical signal is then converted to the decision electric signal by the photodetector. After photodetection, the noise component affecting the electrical signal at the optimum sampling instant can be modeled as a 4-degree of freedom Chi-square random process [7], with variance parameter:

$$\sigma^2 = \frac{N_0}{2} \int_{-\infty}^{+\infty} |H_{RX}(f)|^2 \, df \tag{3}$$

and non-centrality parameter $s^2 = 0$ if $b_n \neq b_{n-1}$ (i.e., $a_n = 0$) and $s^2 = 4P_{S,av}x^2(0)$ if $b_n = b_{n-1}$ (i.e., $a_n = 1$). $H_{RX}(f)$ is the frequency response of the receiver optical filter and N_0 is the one-side power spectral density of ASE noise before optical filtering, that in practical systems is set by the overall amount of noise introduced by the in-line optical amplifiers.

Accordingly to these characteristics of the decision signal and using the theory reported in [9], the expression for the Bit-Error-Rate (BER) for an optical duobinary system can be analytically written as:

$$\mathrm{BER} = \frac{1}{2}\left\{ e^{-\phi}(1 + \phi) + 1 - Q_2\left(\sqrt{\frac{4P_{S,av}x^2(0)}{\sigma^2}}, \sqrt{2\phi} \right)\right\}, \tag{4}$$

where Q_2 is the Marcum Q function of order 2 and ϕ is the decision threshold, that must be optimized for every value of the ratio $P_{S,av}x^2(0)/\sigma^2$. In any case, it can be shown that, independently of the value of ϕ, minimization of BER corresponds to maximization of the first argument of the Marcum Q_2 function. This argument is in fact strictly related to the optical signal-to-noise ratio (OSNR):

$$\frac{4P_{S,av}x^2(0)}{\sigma^2} = 16\,\text{OSNR}\,\frac{\frac{x^2(0)}{T}}{\int_{-\infty}^{+\infty}|H_{RX}(f)|^2\,df} \tag{5}$$

where

$$\text{OSNR} = \frac{P_{S,av}}{2N_0R_B} \tag{6}$$

3.   PULSE SHAPE DEPENDENCE OF BER

The analytical expression of the BER of optical duobinary is similar to that of IMDD [7], except now the first argument of the Q_2 depends on the pulse $x(t)$. This result means that, contrary to IMDD, for a given OSNR, different duobinary pulses may yield different BERs.

To appreciate this, we first assume the transmitted pulse $u(t)$ to be a rectangular pulse of duration T, i.e., the simplest and most typical NRZ pulse. $x(t)$ turns out to be a triangular pulse: $x(t) = 1 - |t/T - 1/2|$ for $t \in [-T/2, 3T/2]$ and $x(t) = 0$ for t outside $[-T/2, 3T/2]$. We get $x(0)/x(T/2) = 1/2$ which, by comparing it to the results presented in [7], shows that there is a penalty with respect to IMDD of exactly 3 dB.

We then select the duobinary pulse with the smallest possible bandwidth occupation [9, 11]:

$$x(t) = \frac{\cos\left(\pi\left[\frac{t}{T} - \frac{1}{2}\right]\right)}{\pi\frac{t}{T}\left(1 - \frac{t}{T}\right)} \tag{7}$$

Now we have $x(0)/x(T/2) = \pi/4$ and the resulting OSNR for BER= 10^{-9} is 16.2, or 12.09 dB, with a *gain* with respect to IMDD of 0.91 dB. This result sets a new quantum limit of 32.4 photons per bit for a conventional optical direct-detection RX.

Between the two considered pulses there is a penalty of almost 4 dB, which shows that the choice of pulse shape is very critical for duobinary. At present, we have not been able to prove that the pulse yielding the lowest possible BER is (7), though we have not been able to find a better performing pulse either.

As a general consideration, we can say that, for any value of OSNR, the best pulse shape $u(t)$ and the best optical filter shape $H_{RX}(f)$ are a unique couple

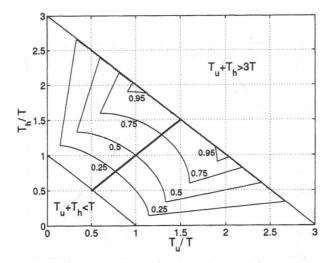

Figure 2. Contour plot of K as a function of both the normalized pulse duration of the transmitted pulse (T_u/T) and of the receiver optical filter impulse response (T_h/T)

and are the ones which maximize the ratio:

$$K = \frac{\frac{2x^2(0)}{T}}{\int_{-\infty}^{+\infty} |H_{RX}(f)|^2 df} = \frac{\frac{2}{T}\int_{-\infty}^{+\infty} u(t) h_{RX}(T/2 - t) dt}{\int_{-\infty}^{+\infty} |h_{RX}(t)|^2 dt} \quad (8)$$

It means that, unlike what happens in standard IMDD systems [10], the optimum receiver for duobinary is based on the pulse-filter pair that maximizes the parameter K.

In order to demonstrate that, in general, the optimum filter is not the one matched to the transmitted pulse shape, we have analyzed the behavior of the parameter K in a simple scenario for which analytical evaluations are straightforward. We assumed that both the transmitted pulse and the receiver optical filter impulse response have a rectangular shape with duration T_u and T_h, respectively. It is important to remark that, in order to comply with the duobinary ISI condition previously reported, T_u and T_h must satisfy the following two constraints [9]:

1. $T_u + T_h \leq 3T$ (otherwise $x(nT) \neq 0$ for some $n \neq 0, 1$);
2. $T_u + T_h > T$ (otherwise $x(0) = x(T) = 0$).

For each possible pair (T_u, T_h), the value of K has been analytically evaluated for the considered scenario. Fig. 2 shows the contour plot of the parameter K as a function of the normalized duration of the transmitted pulse T_u/T and of the receiver optical filter impulse response T_h/T. Regions where $T_u + T_h < T$ and $T_u + T_h > 3T$ do not satisfy the duobinary ISI condition.

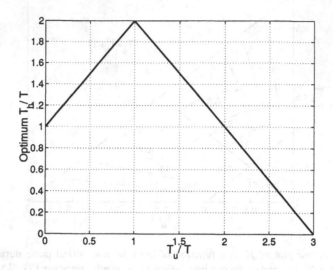

Figure 3.   Plot of the optimum normalized impulse response duration (T_h/T) as a function
the normalized duration of the transmitted pulse(T_u/T).

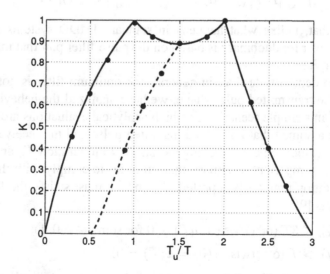

Figure 4.   Plot of the optimum value of the parameter K as a function of T_u/T (solid line).
Dotted line refers to the matched filter (sub-optimum) condition. Results reported as black dots
are obtained through numerical simulation based on error counting.

The thick solid line corresponds to the case of optical filter matched to the transmitted pulse. Maximum values of K, i.e., optimal configurations, correspond to the two pairs $(T_u/T = 1, T_h/T = 2)$ and $(T_u/T = 2, T_h/T = 1)$, which do not belong to the matched filter category. It demonstrates that, in general, the optical matched filter is not the optimum for optical systems using duobinary line-coding. Similar counterexamples can be derived for other pulse and filter shapes.

Fig. 3 and Fig. 4 show as solid lines the optimum normalized filter duration T_h/T and the optimum value of K as a function of T_u/T, respectively. In Fig. 4, the dotted line refers to the matched filter condition: it can be noted that whenever a matched filter setup is a possible choice (i.e., in case $0.5 < T_u/T < 1.5$ so that the duobinary ISI condition is satisfied) there is always a better filtering option based on a narrower filter (longer impulse response duration).

As further verification, numerical simulations based on brute-force error-counting have been carried out: results are shown in Fig. 4 through black dots. A perfect agreement with the analytical results confirms the previous statements.

4. CONCLUSIONS

We have shown that for optical transmission systems based on duobinary line-coding, in general the optimum receiver is not based on the optical filter matched to transmitted pulse-shape. In general, the receiver optical filter must be optimized for each transmitted pulse within the ISI conditions imposed by the duobinary line-coding. In order to achieve such a result, we have derived the expression of the parameter K to be maximized with the purpose to decide the optimal filter for each pulse-shape.

REFERENCES

[1] A. Lender, "The Duobinary technique for high-speed data transmission," *IEEE Trans. Commun. Technol.*, vol. 82, pp. 214–218, May 1963.

[2] K. Yonenaga *et al.*, "Optical Duobinary transmission system with no receiver sensitivity degradation," *Electron. Lett.*, vol. 3, pp. 302–304, Feb. 1995.

[3] X. Gu and L. C. Blank, "10 Gbit/s unrepeated optical transmission over 100 km of standard fibre," *Electron. Lett.*, vol. 29, pp. 2209–2211, 9 Dec. 1993.

[4] D. Penninckx *et al.*, "The Phase-shaped Binary Transmission (PSBT): a new technique to transmit far beyond the chromatic dispersion limit," *Proc. ECOC '96*, Oslo, vol. 2, pp. 173–176.

[5] H. Bissesur *et al.*, "3.2 Tb/s (80 × 40 Gb/s) C-band transmission over 3 × 100 km with 0.8 bit/s/Hz efficiency," in *Proc. ECOC'01*, Amsterdam, The Netherlands, 2001, Paper PD.M.1.11

[6] T. Ono *et al.*, "Characteristics of optical Duobinary signals in terabit/s capacity, high-spectral efficiency WDM systems," *J. Lightwave Technol.*, vol. 16, pp. 788–797, May 1998.

[7]   G. Bosco, A. Carena, V. Curri, R. Gaudino, and P. Poggiolini, "Quantum limit of direct detection optically preamplified receivers using duobinary transmission", *IEEE Photon. Technol. Lett.*, vol. 15, no. 1, pp. 102–104, Jan. 2003.

[8]   P. J. Winzer, "Optically preamplified receiver with low quantum limit," *Electron. Lett.*, vol. 37, no. 9, p. 582, Apr. 2001.

[9]   J. G. Proakis, *Digital Communication*, New York: McGraw-Hill, 1989.

[10] L. Kazovsky *et al.*, *Optical Fiber Communication Systems*, Artech House Inc., 1996.

[11] E. R. Kretzmer, "Generalization of a technique for binary data communication," *IEEE Trans. Commun. Technol.*, vol. COM-14, pp. 67–68, Feb. 1966.

[12] P. Kabal and S. Pasupathy, "Partial response signaling," *IEEE Trans. Commun.*, vol. COM-23, no. 9, pp. 921–934, Sep. 1975.

THEORETICAL LIMITS FOR THE DISPERSION LIMITED OPTICAL CHANNEL

Roberto Gaudino
Politecnico di Torino, Photonlab, Dipartimento di Elettronica, Corso Duca degli Abruzzi, 24, 10129 Torino, Italy
gaudino@polito.it

Abstract: In this paper, we study the theoretical limits of optical communication channels affected by chromatic dispersion. By using as a metric the energy transfer ratio, we find the optimal transmitted pulse shape that allows minimizing the impact of dispersion, together with an interesting definition of the dispersive channel equivalent bandwidth. This paper, though mainly theoretical, tries to approach using a rigorous formalism a problem that is currently receiving large interest, i.e., the optimization of the transmitter for a dispersion-limited optical system.

Key words: optical fiber communication; chromatic dispersion; intersymbol interference; energy transfer ratio.

1. INTRODUCTION

A large amount of theoretical and experimental work has recently focused on finding efficient modulation formats for the so-called "dispersion-limited" optical channel [1], i.e., for an optical link mainly limited by fiber chromatic dispersion. In this paper, we try to approach the problem using a rigorous theoretical formalism, by solving, under a suitable metric discussed below, the problem of the optimization of the transmitted input pulse which leads to the minimization of intersymbol interference (ISI) at the receiver. The paper is mainly theoretical, and allows to define an interesting equivalent bandwidth for the disperion-limited channel. Anyway, it is also open to practical application, as mentioned in a previous paper [2].

2.   MATHEMATICAL ASSESSMENT OF THE PROBLEM

Being interested in dispersion-limited systems, we focus on a fiber transmission model which includes first order chromatic dispersion only, neglecting all other transmission impairments. Thus, we consider the well-known chromatic dispersion transfer function [1]:

$$H_F(f) = e^{-j\frac{\beta_2}{2}L(2\pi f)^2} \tag{1}$$

where $\beta_2 = -\dfrac{\lambda_0 D}{2\pi f_0}$ is the *chromatic dispersion parameter*, being D the *fiber chromatic dispersion* (usually expressed in ps/nm/km), and λ_0 and f_0 the laser central wavelength and frequency, respectively, while L is the fiber length. As commonly accepted, we will indicate as *accumulated dispersion* the quantity $\beta_2 L$ in ps^2 or equivalently DL in ps/nm. We neglect higher order dispersion, such as β_3.

In order to simplify the expressions, we introduce the *Normalized Dispersion Index* (NDI) γ,[1] defined as:

$$\gamma = -2\beta_2 L R^2 \tag{2}$$

where $R = 1/T_B$ is the system bit rate (being T_B the bit duration). The γ parameter is quite useful in simplifying the equations, normalizing them to the system bit rate R or the bit duration T_B. In fact, using this notation, the transfer function and inpulse response become [1]:

$$H_F(f) = e^{j\gamma\left(\frac{\pi f}{R}\right)^2} \rightarrow h_F(t) = \frac{e^{j\pi/4\,\text{sign}(\gamma)}}{T_B\sqrt{\pi|\gamma|}}e^{-\frac{j}{\gamma}\left(\frac{t}{T_B}\right)^2}. \tag{3}$$

so that both time and frequency can be normalized to the bit rate R and bit duration T_B. We assume that the transmitted binary digital signal (complex envelope optical field) is in the form: $x(t) = \sum_{k=-\infty}^{+\infty} \alpha_k \cdot s_{in}(t - kT_B)$ where $s_{in}(t)$ is the complex envelope of the transmitted pulse for a single bit, and α_k assumes the values 0 and 1 for a standard OOK modulation. Since the channel is linear and time invariant (LTI), the resulting pulse at the fiber output is $s_{out}(t) = s_{in}(t) * h_F(t)$. The goal of our paper is the optimization of the input pulse $s_{in}(t)$ under the following assumptions:

- The input pulse $s_{in}(t)$ is strictly time-limited to the interval:

$$I = [-T_{in}/2, +T_{in}/2] \tag{4}$$

---

[1] The γ parameter has already been used by other authors, such as [1], this γ should not be confused with the optical fiber nonlinear Kerr coefficient.

As a particular case, we have $T_{in} = T_B$ for a standard memoryless transmitter, but we will show that the case $T_{in} > T_B$, corresponding to a transmitter with memory, is extremely interesting in extending the dispersion limit. In particular, we will assume $T_{in} = n_{mem} \cdot T_B$, where the (integer) parameter n_{mem} is the transmitter memory. The extension of our results to a modulation with memory is the main new result of this paper with respect to our previous paper on the same topic [2]. Using modulation with memory, the pulse transmitted for each individual bit has a duration that extends beyond the one bit window, an approach that is typical in line coding, such as duobinary [1]. As a practical example, a system working at 10 Gbit/s ($T_B = 100$ ps) with $n_{mem} = 4$ will use pulses at the transmitter side with a duration $T_{in} = 400$ ps. The particular case $n_{mem} = 1$ corresponds to a standard, memoryless modulation. For $n_{mem} > 1$, it should be noted that the signal coming out of the transmitter is affected by ISI. Anyway, line coding is usually associated with a propagation channel that, under suitable conditions, reduces or cancels the amount of ISI present at the transmitter. For instance, optical duobinary can be interpreted as line coding with $n_{mem} = 2$. In fact, the resulting duobinary signal at the transmitter output is strongly affected by ISI, giving rise to a 3-level eye diagram. In the duobinary case, the ISI at the transmitter is anyway cancelled by the direct-detection receiver, which converts the 3-level ISI-affected signal into a standard 2-level signal without ISI. In general, in line coding or modulation with memory, a controlled amount of ISI is created at the transmitter in order to have some kind of advantage at the receiver.

- We chose as optimization criterion the maximization of the energy transfer from the input time window I to an output time window:

$$J = [-T_{out}/2, +T_{out}/2] \qquad (5)$$

More specifically, we introduce the input and output energies:

$$\mathcal{E}_{in} = \int_I |s_{in}(t)|^2 \, dt \quad \text{and} \quad \mathcal{E}_{out} = \int_J |s_{out}(t)|^2 \, dt \qquad (6)$$

and we maximize over $s_{in}(t)$ the *Energy Transfer Ratio* (ETR), defined as:

$$\text{ETR} = \frac{\mathcal{E}_{out}}{\mathcal{E}_{in}} . \qquad (7)$$

The pulses $s_{in}(t)$ resulting from the optimization process proposed here will be indicated as "optimal pulses" in the rest of the paper.

- The criterion we have chosen is particularly relevant for the case $T_{out} = T_B$, if we assume symbol-by-symbol detection for a binary memoryless receiver (i.e., a receiver taking decision on single received bits). The concentration of the output pulse energy over a T_B time window is effective in both minimizing ISI (which is the goal of our paper) and in increasing the signal to noise ratio at the decision instant for any "reasonable" digital receiver. In fact, the criterion is "exact" for an ideal optical integrate&dump receiver, since in this case the decision sample is directly proportional to the signal energy over a T_B time window. Anyway, as we we have shown in [2], it proves a extremely good criterion for realistic optical receiver structures. We notice that, for $n_{mem} > 1$, we are considering a somehow non-intuitive system where ISI is strongly present at the transmitter side, but then ISI is reduced, or even cancelled at the receiver side by the propagation over the dispersive channel.

- We will showed in [2] that the ETR (for $T_{out} = T_B$) for realistic optical receivers should typically be above 90% to give a penalty due to ISI in the 1-2 dB range. As a consequence, we will conventionally define in the rest of the paper the "dispersion limit" as the amount of accumulated dispersion for a given bit rate that results in an ETR = 90%.

2.1   The fundamental parameters and equations

The ETR optimization problem over a generic LTI system is a canonical problem that was studied in the past [3, 4] and can be reduced to the optimization of a quadratic functional in $s_{in}(t)$, with a quadratic constraint. For a generic filter impulse response, it leads to the following Fredholm integral equation of the second kind:

$$\int_I \mathcal{K}(u,v)\, s_{in}(u)\, du = \lambda\, s_{in}(v) \tag{8}$$

where the kernel of the integral equation is:

$$\mathcal{K}(u,v) = \int_J h_F(z-u)\, h_F^*(z-v)\, dz \tag{9}$$

and where the optimal solution is given by the eigenfunction corresponding to the *maximum eigenvalue* λ, which is equal to the ETR (7) [3].

This problem has been solved in the literature for several types of band-limited and standard low-pass filters [3–5]. In this paper, we solve it (for the first time to our knowledge) considering the fiber dispersive transfer function (1) as the band-limiting filter. In this case, replacing (3) in (9), by straigthforward calculations, the kernel can be expressed as:

$$\mathcal{K}(u,v) = \frac{e^{-j\left(\frac{u^2 - v^2}{\gamma\, T_B^2}\right)}}{\pi(u-v)} \cdot \sin\left[\frac{T_{out}(u-v)}{|\gamma|\, T_B^2}\right]. \tag{10}$$

2.2   The closed form solution

The integral equation (8), with the kernel (10), can be solved by looking for a solution in the form $s_{in}(t) = a(t) \cdot e^{j\phi(t)}$, where $a(t)$ and $\phi(t)$ are real functions of time. This separates the two input pulse contributions that are usually called amplitude modulation and phase modulation (or chirp). In particular, we look for a solution of the form:

$$s_{in}(t) = a(t) \cdot e^{\frac{j}{\gamma}\left(\frac{t}{T_B}\right)^2}. \tag{11}$$

This "guess" was originally driven by the observation of the numerical results obtained in [2], and proved to be exact, as shown in the following. By writing the kernel as $\mathcal{K}(u,v) = \mathcal{K}_{\mathcal{R}}(u,v) \cdot \exp\left[-j\left(\frac{u^2-v^2}{\gamma T_B^2}\right)\right]$, where:

$$\mathcal{K}_{\mathcal{R}}(u,v) = \frac{1}{\pi(u-v)} \cdot \sin\left[\frac{T_{out}(u-v)}{|\gamma|T_B^2}\right] \tag{12}$$

and by substituting (11) into (8), the phase terms vanish and the resulting integral equation simplifies to:

$$\int_I \mathcal{K}_{\mathcal{R}}(u,v)\,a(u)\,du = \lambda\,a(v) \tag{13}$$

This is thus the fundamental integral equation that allows solving our optimization problem. It turns out that exactly the same integral equation results from the ETR pulse optimization over an ideal low-pass filter with bandwidth W and $J = [-\infty, +\infty]$. This is a very fortunate case, since the ideal low-pass filter problem received a lot of attention in the past, in the framework of fundamental works on communications theory, and it was fully analyzed and analytically solved in [5]. It leads to an integral equation with kernel:

$$\mathcal{K}_{LP}(u,v) = \frac{\sin[2\pi W(u-v)]}{\pi(u-v)}. \tag{14}$$

Thus, the integral equation (13) is mathematically equivalent to the ideal low-pass case. A full treatment of these results can be found in [6], [7], or in [8], where the expression of the result in terms of Prolate Spheroidal functions is given. In particular, it turns out that the solution corresponding to the the maximum eigenvalue of (13) corresponds to the one maximizing the ETR. By direct comparison between the kernels (12) and (14), we observe that we have the following equivalence among parameters:

$$2\pi W = \frac{T_{out}}{|\gamma|\,T_B^2} \quad \Rightarrow \quad W = \frac{T_{out}}{2\pi\,|\gamma|\,T_B^2}. \tag{15}$$

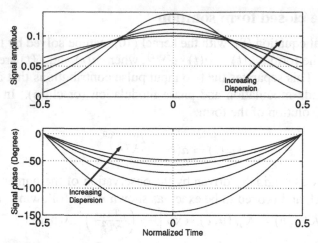

Figure 1. Optimal pulses for $n_{mem} = 1$ and γ values ranging from 0.1 to 0.3 in 0.05 steps. Time is normalized to T_B.

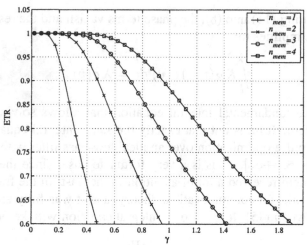

Figure 2. ETR vs. γ for $n_{mem} = 1, 2, 3, 4$.

The availability of a closed form solution is, in our opinion, the most important result of this paper, not only because it gives the optimal pulse expressed through prolate spheroidal functions [5], but even more because it leads to the interesting results we illustrate in the following Section. We show in Fig. 1 the resulting optimal pulses for $n_{mem} = 1$ and γ values ranging from 0.1 to 0.3 in 0.05 steps, while we show in Fig. 2 the ETR vs. γ for different n_{mem} values.

3.   OPTIMAL CHIRP AND CHANNEL EQUIVALENT BANDWIDTH

The previous results lead to the following important considerations:

1. The optimal pulses have a phase modulation given by $\phi(t) = \frac{t^2}{\gamma T_B^2}$, or equivalently, $\phi(t) = -\frac{t^2}{2\beta_2 L}$. This expression gives the optimal chirp for pulses launched over a dispersive channel. Interestingly, this result was already found in [1] using a totally different approach for which anyway optimality was not proven.

2. Provided that the pulse chirp is chosen to be optimal, the dispersive channel is totally equivalent, at least in the ETR sense, to an ideal low-pass filter with bandwidth:

$$W = \frac{T_{out}}{2\pi |\gamma| T_B^2} \quad \text{or equivalently} \quad W = \frac{1}{4\pi} \frac{T_{out}}{|\beta_2| L} . \qquad (16)$$

   This result can be usefully interpreted as a definition of the *equivalent bandwidth* of the dispersive channel which, to our knowledge, was never given before in a rigorous form. We note here that dispersive channel transfer function (1) has a peculiar expression that render most of the common bandwidth definitions totally useless, since $|H_F(f)|^2 = 1, \forall f$. For instance, the commonly used noise equivalent bandwidth is infinite, and the 3-dB bandwidth is meaningless.

3. In the ideal low-pass problem with kernel (14), it can be shown that the ETR depends *only* on WT_{in}, and the function ETR $= f(WT_{in})$ is monotonically increasing, asymptotically reaching ETR $= 1$ for $WT_{in} \to +\infty$ [5]. If we fix to the limiting value ETR $= 0.9$ (a 90% energy transfer), the condition $WT_{in} \geq 0.675$ must be satisfied [5]. In our case, using (16), the ETR is a function of $\frac{T_{out} T_{in}}{2\pi |\gamma| T_B^2}$ only. If we fix $T_{out} = T_B$ and $T_{in} = n_{mem} \cdot T_B$, we have that the ETR is only a function of $n_{mem}/|\gamma|$.

4. In order to have ETR $= 0.9$, for $n_{mem} = 1$, we have the condition $|\gamma| \leq 0.236$, or equivalently, introducing (2),

$$R^2 \leq \frac{0.1179}{|\beta_2 L|} . \qquad (17)$$

   This last equation can be interpreted as the theoretical upper bound to the maximum bit rate that can be achieved over the dispersive optical channel with limited ISI (i.e. ETR $= 0.9$) and for the memoryless modulator ($n_{mem} = 1$).

5. From another point of view, if we need to transmit over a fiber with arbitrary bit rate and dispersion, we can *always* obtain a limited ISI condition (e.g., ETR ≥ 0.9) provided that we accept a memory n_{mem} at the transmitter given by:

$$n_{mem} \geq 4.241 |\gamma| \quad \Rightarrow \quad n_{mem} \geq 8.482 |\beta_2 L| R^2 . \qquad (18)$$

Table 1.   Maximum acceptable accumulated dispersion values (in terms of DL in ps/nm) for 10 and 40 Gbit/s system for a given memory n_{mem}.

n_{mem}	10 Gbit/s	40 Gbit/s
1	928 ps/nm	58 ps/nm
2	1856 ps/nm	116 ps/nm
3	2785 ps/nm	174 ps/nm
4	3713 ps/nm	232 ps/nm

This is a novel and important result, stating that, we can limit ISI provided that the modulator memory n_{mem} is sufficiently large, and optimal pulses are used. Table 1 reports the amount of accumulated dispersion that, according to (18), can be tolerated for a 10 and 40 Gbit/s system for different n_{mem}.

The result expressed in (18) also states that the dispersive channel, for arbitrary values of dispersion, allows an arbitrarily high bit-rate, provided that n_{mem} is sufficiently large and, obviously, that optimal pulses are used. Practically, this means that for high dispersion, the optimal pulses are compressed by the channel from an input duration $n_{mem} \cdot T_B$ to an output duration close to $T_{out} = T_B$.

4.   CONCLUSION

We have approached the problem of the optimization of the transmitted pulse in a dispersion-limited optical channel using a rigorous approach. We believe that our work, though mainly theoretical, can give a useful insight on the problem on optical line coding, and be a good complement of the approach followed in papers such as [1].

REFERENCES

[1] E. Forestieri and G. Prati, "Novel optical line codes tolerant to fiber chromatic dispersion," *J. Lightwave Technol.*, vol. 19, no. 11, pp. 1675–1684, Nov. 2001.

[2] R. Gaudino and E. Viterbo, "Pulse shape optimization in dispersion-limited direct detection optical fiber links," to appear on *IEEE Comm. Lett.*

[3] M. Elia, G. Taricco, and E. Viterbo, "Optimal energy transfer in band-limited communication channels," *IEEE Trans. Inform. Theory*, vol. 45, no. 6, pp. 2020–2029, Sept. 1999.

[4] L. E. Franks, *Signal Theory*, Prentice-Hall, 1969.

[5] D. Slepian, "Some comments on Fourier analysis, uncertainty and modeling," *SIAM Rev.*, vol. 25, no. 3, pp. 379–393, July 1982.

[6] D. Slepian and H. Pollak, "Prolate spheroidal wave functions, Fourier analysis and uncertiainty - I", *Bell System Tech. J.*, vol. 40, pp. 43–63, Jan. 1961.

[7] D. Slepian and H. Pollak, "Prolate spheroidal wave functions, Fourier analysis and uncertiainty - II", *Bell System Tech. J.*, vol. 40, pp. 65–84, Jan. 1961.

[8] *Data communication: fundamentals of baseband transmission,* Benchmark papers in Electrical Engineering and computer science, vol. 9, Halsted Press, 1974.

CAPACITY BOUNDS FOR MIMO POISSON CHANNELS WITH INTER-SYMBOL INTERFERENCE

Alfonso Martinez

Technische Universiteit Eindhoven; Electrical Eng. Dept., Signal Processing Group (SPS); Den Dolech 2, P. O. Box 513, 5600 MB Eindhoven, The Netherlands

alfonso.martinez@ieee.org

Abstract: We study multiple-access Poisson channels with multiple receivers. Each transmitter sends a sequence of modulated symbols, which may also be affected by inter-symbol interference. We derive some new formulas for the channel capacity, for the cases of both independent and coordinated transmission. We also provide some numerical results on the additional power required for efficient transmission due to the various sources of interference.

Key words: channel capacity; MIMO systems; intersymbol interference.

1. INTRODUCTION

A Poisson channel models an optical communication channel in which the fundamental impairment is shot noise. A signal is observed through a sequence of arrivals in a detector, and the arrivals follow a (random) Poisson process. Their position, which we also call arrival instants, counts, or time stamps, are known with arbitrarily good precision.

We shall study multiple-input multiple-output (MIMO) Poisson channels. A total of L signals are transmitted, denoted by $\lambda_l^T(t)$, with $1 \leq l \leq L$. These signals experience static mixing with coefficients by $H_{l,r}, 1 \leq l \leq L, 1 \leq r \leq R$; the mixing coefficients are real positive numbers, $H_{l,r} \geq 0$. The signals are detected by an array of R elements, each of them detecting a signal $\lambda_r(t) = \sum_{l=1}^L H_{l,r}\lambda_l^T(t)$. The arrival times at receiver r are denoted by $\underline{\tau}_r$.

Sect. 2 presents the capacity of an unconstrained MIMO channel, which corresponds to an ideal, non-dispersive, infinite-bandwidth system.

Some constraints typical of a practical optical channel are imposed in Section 3, where closed-form bounds to the capacity are presented and computed. Following standard practice, the signals are modulated with On-Off Keying (OOK), and are given by:

$$\lambda_i^T(t) = \sum_{n=0}^{N-1} A_i w_i(n) h(t - nT - \Delta_i). \tag{1}$$

With no loss of generality, we assume the following: A_i is the energy for the "on" symbol; $w_i(n) \in \{0,1\}$ are the modulation symbols; $h(t)$ in the modulation pulse, real-valued, positive and of area 1; Δ_i is the relative delay of channel i; a total of N symbols are transmitted. Note that dispersion is modelled as inter-symbol interference. Fig. 1 depicts the sequences transmitted in a system with two input channels.

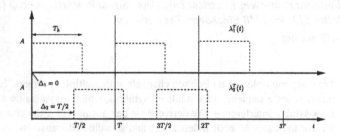

Figure 1. Transmitted signals: 2 OOK parallel channels.

2. UNCONSTRAINED MIMO POISSON CHANNEL

We extend the formula for single-input single-output (SISO) Poisson channel [1] with the following:

THEOREM 1: *For a Poisson channel with R receiver lines, the mutual information between the input intensity $\lambda(t) = \big((\lambda_1(t),\ldots,\lambda_R(t)\big)$, and the output $\underline{\tau} = (\tau_1,\ldots,\tau_R)$ is given by:*

$$I(\lambda(t); \underline{\tau}) = \sum_{r=1}^{R} \int E_{\lambda_r(t)}\big\{\psi(\lambda_r(t))\big\} - E_{I_\alpha(t,r)}\big\{\psi(\hat{\lambda}_r(t))\big\}\, dt, \tag{2}$$

where $\psi(x) \triangleq x\log(x)$, $E_{I_\alpha(t,r)}\{\cdot\}$ is the expectation over a past $I_\alpha(t,r)$ from instant t seen at receiver r, and $\hat{\lambda}_r(t)$ is defined as the expected value of the projection from the past $I_\alpha(t,r)$.

Note that the past $I_\alpha(t,r)$ is arbitrary, and need not be the causal ordering of time. The proof can be found in Appendix 2. The proof is constructive, and also provides a capacity-achieving input distribution. Similar to the SISO channel (see Kabanov [1], Davis [2] and Wyner [3, 4]), a simple two-valued signal, such that all L signals are simultaneously active or inactive, is enough to achieve capacity. The distribution of the on and off states can be found in [3, 4]. If the signal duration is made ever smaller, all available degrees of freedom are used.

3. CONSTRAINED MIMO POISSON CHANNELS

In presence of modulation, we calculate the capacity per channel use, having defined a channel use as a symbol interval of duration T:

$$C = \sup_{\lambda(t)} \lim_{N \to +\infty} \frac{1}{N} I(\lambda(t); \underline{\tau}). \tag{3}$$

In practice, the limit will be truncated to a small value of N, and we shall assume that convergence in N has been achieved.

Even though (2) is a closed formula, it is not easily applicable to situations with bandwidth limitations, inter-symbol interference, and dispersion. The expectation operator $E_{I_\alpha(t,r)}\{\cdot\}$ is not easily tractable. Lapidoth and Moser [5] have recently calculated upper and lower bounds for the SISO discrete-time Poisson channel. They approximate the differential entropies in Eq. (4) by using some properties of Poisson rv. We are, however, interested in the continuous-time channel, for which their approximations fail.

An alternative route starts at the expression with the differential entropies (for a proof, see Appendix 2):

$$I(\lambda(t); \underline{\tau}) = h(\underline{\tau}) - h(\underline{\tau}|\lambda(t)), \tag{4}$$

$$= h(\underline{\tau}) + \sum_{r=1}^{R} \int E_{\lambda_r(t)} \{\psi(\lambda_r(t))\} dt - \sum_{r=1}^{R} \int E_{\lambda_r(t)} \{\lambda_r(t)\} dt. \tag{5}$$

Our calculations of the channel capacity will nevertheless involve several approximations:

- All symbols $w_i(n)$ are iid (no space-time coding). This is equivalent to having equiprobable signals $\lambda(t)$.

- Instead of the exact arrival times $\underline{\tau}$, we use the number of arrivals. We partition the time axis in disjoint intervals, and count the number of arrivals in each, which we denote by k_i, and $\underline{k}$ for the whole partition. Due to the data processing inequality this may in general decrease the mutual information, as there may be information in the arrival instants. Appendix 2 elaborates on the relationship between the differential entropy $h(\underline{\tau})$ and the entropy $H(\underline{k})$, and the conditions under which they are equivalent.

Taken together, we obtain a lower bound to the channel capacity. Furthermore, as we are interested in symmetric situations, all users are assumed identical;

this implies in particular that there information rates and energies are identical. A simple upper bound is then obtained by considering the single-user capacity: the presence of other users cannot increase the average amount of information that can be transmitted.

Fig. 2 shows the capacity, measured in bits per symbol period (T), for an OOK scheme, and estimated with the equations presented above. The pulses have a length equal to T_h, as indicated in the plot, so that they may overlap with each other, a model for dispersion. The capacity is estimated in bits per symbol period, so that the maximum is 2, as there are two input channels. The channel matrix has a flat spatial response, $H_{l,r} = 0.5, l, r = 1, 2$. The second user has a delay $\Delta_2 = T/2$. Note that the model also corresponds to a situation with a single user corrupted by ISI. For comparison purposes, we also report the unconstrained capacity, and show the losses incurred by binary modulation, and then by multiple-access effects.

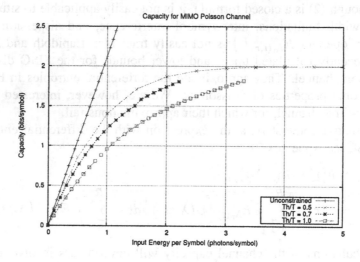

Figure 2.   Capacity with OOK, from exact calculations and from simulations.

APPENDIX

The proof presented here is based on simple information-theoretic arguments, and as such differs from previous ones [1], which made use of martingales.

We start by stating two lemmas that will be used later. The proofs are sketchy, thanks to their simplicity.

LEMMA 2: *Let n be a Poisson random variable with parameter λ, where λ is itself randomly distributed in an interval $[0, A]$, $A \to 0$, with a density $p(\lambda)$. The conditional entropy $H(n|\lambda)$ is given by:*

$$H(n|\lambda) = \int_\lambda p(\lambda)(\lambda - \lambda \log \lambda) d\lambda = E\lambda - E\psi(\lambda) \tag{A.1}$$

with $E\lambda = \int_0^A \lambda p(\lambda)\, d\lambda$, and $\psi(x) = x \log x$.

PROOF: As $A \to 0$, we can safely disregard second-order terms in A. The Poisson distribution tends to a Bernoulli variable, with only two possible values, 0 and 1. From the definition of entropy, in the limit $A \to 0$, and disregarding second order terms we obtain the desired expression. ∎

LEMMA 3: *Let n be a Poisson random variable with parameter λ, where λ is itself randomly distributed in an interval $[0, A]$, $A \to 0$, with a density $p(\lambda)$. The entropy $H(n)$ is given by:*

$$H(n) = E\lambda - E\lambda \log E\lambda = E\lambda - \psi(E\lambda) \tag{A.2}$$

with $E\lambda = \int_0^A \lambda p(\lambda)\, d\lambda$, and $\psi(x) = x \log x$.

PROOF: The same approximations mentioned in the proof of Lemma 2 are needed here. ∎

Now we are in position of proving Theorem 1:

PROOF: Let us divide the observation interval $I = (t_0, t_0 + T)$ in M disjoint cells of length Δ each. M depends on T and Δ so as to satisfy $M\Delta = T$. Each cell is centered at a point t_m, $1 \le m \le M$. This creates a lattice of $M \times R$ observation cells, M for each of the R receivers. Let us denote a cell with $I(m, r)$, with the obvious meaning for the indices m and r.

1. The input in interval $I(m, r)$ is given by $\int_\Delta \lambda_r(t)\, dt \simeq \lambda_r(t_m)\Delta$. Let us denote the input in the cell $I(m, r)$ by $\lambda_r(t_m)$, and the set of all inputs by $\underline{\lambda} \triangleq \bigcup_{r=1}^R (\lambda_r(t_1), \ldots, \lambda_r(t_M))$.

2. The output is the number of arrivals $n(m, r)$ in $I(m, r)$, and is Poisson distributed with parameter $\lambda_r(t_m)\Delta$. For M large enough, and a fixed t_k, the number of arrivals is Bernoulli distributed along the receivers. Let us define $n(m, r)$ as the number of arrivals in the cell $I(m, r)$, and $\underline{n} \triangleq \bigcup_{r=1}^R (n(1, r), \ldots, n(M, r))$.

The mutual information between the input intensities $\lambda(t)$ and the arrivals $\underline{\tau}$ is now given by:

$$I(\lambda(t); \underline{\tau}) = \lim_{M \to +\infty} I\Big(\bigcup_{r=1}^R \lambda_r(t_1), \ldots, \lambda_r(t_M); \bigcup_{r=1}^R n(r, 1), \ldots, n(r, M)\Big) \tag{A.3}$$

$$= \lim_{M \to +\infty} I(\underline{\lambda}; \underline{n}) = \lim_{M \to +\infty} H(\underline{n}) - H(\underline{n}|\underline{\lambda}) \tag{A.4}$$

Let us start with the conditional entropy $H(\underline{n}|\underline{\lambda})$. For a given intensity λ, the outputs are conditionally independent. We apply Lemma 2 to each of these terms:

$$H(\underline{n}|\underline{\lambda}) = \sum_{r=1}^R \sum_{m=1}^M E_{\lambda_r(t_m)} H\big(n(m, r)|\lambda_r(t_m)\big) \tag{A.5}$$

$$= \sum_{r=1}^R \sum_{m=1}^M \Big(E\{\lambda_r(t_m)\}\Delta - E\{\lambda_r(t_m)\}\psi(\Delta)\Big)$$

$$- \sum_{r=1}^R \sum_{m=1}^M \int_{\lambda_r(t_m)} p(\lambda_r(t_m))\Delta\,\psi(\lambda_r(t_m))\, d\lambda_r(t_m) \tag{A.6}$$

Let us now go back to the term $H(\underline{n})$. We now define the past as the set of all cells coming before $I(m, r)$, and denote it with $I_\alpha(m, r)$[1]. For the time being we leave the ordering arbitrary.

---

[1] Any ordering would be valid, and that natural need not be followed.

We can decompose it as a sum:

$$H(\underline{n}) = \sum_{r=1}^{R} \sum_{m=1}^{M} E_{n(I), I \in I_\alpha(m,r)} H(n(m,r)|n(I)) \tag{A.7}$$

From the memoryless property of the Poisson process, the sequence $(n(I), I \in I_\alpha(m,r)) \to \lambda_r(t_m) \to n(m,r)$ forms a Markov chain, so that we can define an a posteriori probability $p_\alpha(\lambda_r(t_m)|n(I))$[2]:

$$p_\alpha(\lambda_r(t_m)|n(I)) \triangleq p(n(I)|\lambda_r(t_m)), \quad I \in I_\alpha(m,r) \tag{A.8}$$

We now define an equivalent $\hat{\lambda}_r(t_m)$, as the expected value of the estimate of $\lambda_r(t_m)$ from the past, and whose density is given by $p_\alpha(\lambda_r(t_m)|n(I))p(\lambda_r(t_m))$:

$$\hat{\lambda}_r(t_m) = \int_{\lambda_r(t_m)} p(\lambda_r(t_m)) \, p_\alpha(\lambda_r(t_m)|n(I)) \, \lambda_r(t_m) \Delta \, d\lambda_r(t_m) \tag{A.9}$$

We now invoke Lemma 3 to calculate the value of $H(n(m,r)|n(I))$:

$$H(n(m,r)|n(I)) = \hat{\lambda}_r(t_m) - \psi(\hat{\lambda}_r(t_m)) \tag{A.10}$$

This allows us to rewrite Eq. (A.7) in a more convenient form:

$$H(\underline{n}) = \sum_{r=1}^{R} \sum_{m=1}^{M} E_{n(I), I \in I_\alpha(m,r)} H(n(m,r)|n(I)) \tag{A.11}$$

$$= \sum_{r=1}^{R} \sum_{m=1}^{M} E_{n(I), I \in I_\alpha(m,r)} \left(\hat{\lambda}_r(t_m) \Delta - \hat{\lambda}_r(t_m) \psi(\Delta) \right)$$

$$- \sum_{r=1}^{R} \sum_{m=1}^{M} E_{n(I), I \in I_\alpha(m,r)} \psi(\hat{\lambda}_r(t_m)) \Delta \tag{A.12}$$

We now exploit the fact that by construction,

$$E_{n(I), I \in I_\alpha(m,r)} \hat{\lambda}_r(t_m) = \tag{A.13}$$

$$= E_{n(I), I \in I_\alpha(m,r)} \int_{\lambda_r(t_m)} p(\lambda_r(t_m)) \, p_\alpha(\lambda_r(t_m)|n(I)) \, \lambda_r(t_m) \, d\lambda_r(t_m) \tag{A.14}$$

$$= \int_{\lambda_r(t_m)} \left(E_{n(I), I \in I_\alpha(m,r)} p_\alpha(\lambda_r(t_m)|n(I)) \right) p(\lambda_r(t_m)) \, \lambda_r(t_m) \, d\lambda_r(t_m) \tag{A.15}$$

$$= \int_{\lambda_r(t_m)} 1 \cdot p(\lambda_r(t_m)) \, \lambda_r(t_m) \, d\lambda_r(t_m) = E\{\lambda_r(t_m)\}. \tag{A.16}$$

Common terms vanish in Eqs. (A.6) and (A.12), and the mutual information is given by:

$$I(\underline{\lambda};\underline{n}) = \sum_{r=1}^{R} \sum_{m=1}^{M} E_{\lambda_r(t_m)} \left\{ \Delta \, \psi(\lambda_r(t_m)) \right\} - \sum_{r=1}^{R} \sum_{m=1}^{M} E_{n(I), I \in I_\alpha(m,r)} \left\{ \Delta \, \psi(\hat{\lambda}_r(t_m)) \right\}, \tag{A.17}$$

---

[2]Note that if there is no statistical correlation among $\lambda_r(t_m)$, the past will not give extra information on the present, and the a posteriori probability will be irrelevant.

where the last equality follows from the same reasoning as in Eqs. (A.13)–(A.16). Finally, in the limit $M \rightarrow +\infty$, the summation becomes an integral, and we obtain

$$I(\lambda(t); \underline{\tau}) = \sum_{r=1}^{R} \int_{t} E_{\lambda_r(t)} \Big\{ \lambda_r(t) \log(\lambda_r(t)) \Big\} dt - \sum_{r=1}^{R} \int_{t} E_{\underline{\tau}(I), I_\alpha(t,r)} \Big\{ \hat{\lambda}_r(t) \log(\hat{\lambda}_r(t)) \Big\} dt.$$

(A.18)

$I_\alpha(t,r)$ is the continous-time ordering derived from the discrete equivalent $I_\alpha(m,r)$. ∎

APPENDIX B

THEOREM 4: *For a given channel intensity $\lambda(t)$, defined in a time interval $(t_0, t_0 + T)$, the entropy of the observed sequence of arrival times $\underline{\tau}$ is given by:*

$$h(\underline{\tau}|\lambda(t)) = \int_{t_0}^{t_0+T} \lambda(t)\, dt - \int_{t_0}^{t_0+T} \lambda(t) \log \lambda(t)\, dt.$$

(B.1)

PROOF: Starting with the definition of entropy, and using the well-known expression for the probability $\Pr(\underline{\tau}|\lambda(t)) = e^{-\Lambda} \prod_{i=1}^{k} \lambda(\tau_i)$, we get:

$$h(\underline{\tau}|\lambda(t)) = -\sum_{k=0}^{+\infty} \int_{\underline{\tau}} e^{-\Lambda} \prod_{i=1}^{k} \lambda(\tau_i) \log\left(e^{-\Lambda} \prod_{i=1}^{k} \lambda(\tau_i)\right) d\underline{\tau},$$

(B.2)

where we group the terms in the sum so that each $\underline{\tau}$ consists of k arrivals. Let this contribution to the entropy be denoted by h_k. By symmetry, the partition of the interval $(t_0, t_0 + \Delta)^k$ is such that it covers an area exactly $1/k!$ of the total area of the original interval. We thus obtain

$$h_k(\underline{\tau}|\lambda(t)) = -e^{-\Lambda} \int_{\tau_1} \cdots \int_{\tau_k > \tau_{k-1}} \lambda(\tau_1) \cdots \lambda(\tau_k)\left(-\Lambda + \sum_{i=1}^{k} \log \lambda(\tau_i)\right) d\tau_1 \cdots d\tau_k$$

$$= -e^{-\Lambda} \frac{1}{k!} \int_{\tau_1} \cdots \int_{\tau_k} \lambda(\tau_1) \cdots \lambda(\tau_k)\left(-\Lambda + \sum_{i=1}^{k} \log \lambda(\tau_i)\right) d\tau_1 \cdots d\tau_k.$$

(B.3)

After some more calculations, we obtain

$$h_k = e^{-\Lambda} \frac{1}{k!} \Lambda \left(\int_{\tau} \lambda(\tau)\, d\tau\right)^k - e^{-\Lambda} \frac{1}{k!} \sum_{i=1}^{k} \int_{\tau_1} \cdots \int_{\tau_k} \log \lambda(\tau_i) \lambda(\tau_1) \cdots \lambda(\tau_k)\, d\tau_1 \cdots d\tau_k$$

$$= e^{-\Lambda} \frac{1}{k!} \Lambda^{k+1} - e^{-\Lambda} \frac{1}{k!} k\left(\int_{\tau} \lambda(\tau)\, d\tau\right)^{k-1} \left(\int_{\tau} \lambda(\tau) \log \lambda(\tau)\, d\tau\right)$$

(B.4)

$$= e^{-\Lambda} \frac{1}{k!} \Lambda^{k+1} - e^{-\Lambda} \frac{1}{(k-1)!} \Lambda^{k-1} \Lambda_L.$$

(B.5)

where $\Lambda = \int_{\tau} \lambda(\tau)\, d\tau$ and $\Lambda_L \triangleq \int_{\tau} \lambda(\tau) \log \lambda(\tau)\, d\tau$. Putting all terms together:

$$h(\underline{\tau}|\lambda(t)) = \sum_{k=0}^{+\infty} h_k(\underline{\tau}|\lambda(t)) = \sum_{k=0}^{+\infty} e^{-\Lambda} \frac{1}{k!} \Lambda \Lambda^k - \sum_{k=1}^{+\infty} e^{-\Lambda} \frac{1}{(k-1)!} \Lambda^{k-1} \Lambda_L$$

$$= \Lambda e^{-\Lambda} e^{\Lambda} - \Lambda_L e^{-\Lambda} e^{\Lambda} = \Lambda - \Lambda_L.$$

(B.6)

∎

APPENDIX C

PROPOSITION 5: *Let the signal intensity $\lambda(t)$ be piecewise-constant, with the points of change (discontinuity) denoted by $t_i, i = 0, \ldots, \ell$, ℓ may be infinite for a countable number of discontinuities. Each interval is then of the form $[t_i, t_{i+1})$; let us define $\Delta t_i \triangleq t_{i+1} - t_i$. Let $\underline{k}$ denote a vector with the number of arrivals at each interval, and $\underline{\tau}$ the corresponding arrival times. The probabilities $\Pr(\underline{\tau}|\lambda(t))$ and $\Pr(\underline{k}|\lambda(t))$ are linked via the following equation:*

$$\Pr(\underline{\tau}|\lambda(t)) = \Pr(\underline{k}|\lambda(t)) \prod_{i=0}^{\ell} \frac{k_i!}{\Delta t_i^{k_i}}. \tag{C.1}$$

PROOF: The signals $\lambda(t)$ are constant in each interval $[t_i, t_{i+1})$, so that the $\lambda(\tau_j)$ is a function of the interval number only. If we now integrate over all possible arrival sequences compatible with a given vector $\underline{k}$, we obtain

$$\Pr(\underline{k}|\lambda(t)) = \int_{\underline{\tau}} e^{-\Lambda} \prod_{i=1}^{\ell} \prod_{j=1}^{k_i} \lambda(\tau_j) \, d\underline{\tau} = e^{-\Lambda} \prod_{i=1}^{\ell} \frac{\prod_{j=1}^{k_i} \lambda(\tau_j)(\Delta t_i)}{k_i!}. \tag{C.2}$$

Using the constantness property, we obtain the desired equation by grouping terms. ∎

PROPOSITION 6: *Under the assumptions of Proposition 5 the entropies $h(\underline{\tau})$ and $H(\underline{k})$ are related via the following equation:*

$$h(\underline{\tau}) = -\sum_{k} \Pr(\underline{k}) \log\left(\Pr(\underline{k}) \prod_{i=0}^{\ell} \frac{k_i!}{\Delta t_i^{k_i}} \right) \tag{C.3}$$

$$= H(\underline{k}) - \sum_{k} \Pr(\underline{k}) \sum_{i=0}^{\ell} \log \frac{k_i!}{\Delta t_i^{k_i}}. \tag{C.4}$$

PROOF: Note that the extra factor in Eq. (C.1) does not depend on the signal $\lambda(t)$, but only on the instants t_i, so that the same constant of proportionality holds for total probabilities $\Pr(\underline{\tau}) = \sum_{\lambda} \Pr(\underline{\tau}|\lambda(t))$.

We can again decompose the integration over $\underline{\tau}$ in terms for fixed $\underline{k}$. For fixed $\underline{k}$, all the compatible terms $\tau^{(k)}$ do not depend on time, and the integral of $\Pr(\tau^{(k)}) \log \Pr(\tau^{(k)})$ is very easy, that of a constant. We now exploit that this integral gives the same proportionality term as Proposition 5, and simple calculations yield now Eq. (C.4). ∎

REFERENCES

[1] Y. Kabanov, "The capacity of a channel of the Poisson type," *Theory of Probability and Applications*, vol. 23, pp. 143–147, 1978.

[2] M. H. A. Davis, "Capacity and cutoff rate for Poisson-type channels," *IEEE Trans. Inform. Theory*, vol. 26, pp. 710–714, November 1980.

[3] A. D. Wyner, "Capacity and error exponent for the direct detection photon channel - part I," *IEEE Trans. Inform. Theory*, vol. 34, pp. 1449–1461, November 1988.

[4] A. D. Wyner, "Capacity and error exponent for the direct detection photon channel - part II," *IEEE Trans. Inform. Theory*, vol. 34, pp. 1462–1471, November 1988.

[5] A. Lapidoth and S. M. Moser, "Bounds on the capacity of the discrete-time Poisson channel," in *Proceedings of the 41st Allerton Conf. on Communication, Control, and Computing*, October 2003.

QSPACE PROJECT: QUANTUM CRYPTOGRAPHY IN SPACE

Cesare Barbieri[1], Gianfranco Cariolaro[2], Tommaso Occhipinti[2],
Claudio Pernechele[3], Fabrizio Tamburini[1,2], and Paolo Villoresi[2]

[1]*Dept. of Astronomy, University of Padova, vicolo dell'Osservatorio 2, I-35122 Padova, Italy;*
[2]*Department of Information Engenieering, University of Padova, via Gradenigo 6/B, I-35131 Padova, Italy;*
[3]*INAF, Cagliari Observatory, Strada 54, Poggio dei Pini I-09012 Capoterra (CA), Italy.*

Abstract: The purpose of this project is to improve the techniques of Quantum Cryptography, to realize a Quantum Key Distribution (QKD) in free space with an orbiting satellite using nowadays technology. With this experiment we characterize the properties of a single photon communication channel from an orbiting satellite in space to a ground based station. We used the facilities of the ASI laser ranging station MLRO (Matera) and the satellite for geodesy Lageos I, equipped with corner-cube retroreflectors, to simulate a single photon transmission from an orbiting satellite.

Key words: optical communication; quantum cryptography; quantum communication; space technology.

1. INTRODUCTION

Quantum mechanics provides powerful tools that form one of the cornerstones of scientific progress, and which are indispensable for nowadays technology. The most important areas where the applications of Quantum mechanics will be crucial in the next future are the new developments in modern communication and information-processing technologies, namely Quantum Communication, Quantum Teleportation with entangled states and Quantum Computation. Quantum Cryptography is the most promising application of Quantum Communication in every day's life of the next future, together with Quantum Teleportation [1–4]. Here the fundamental properties of quantum mechanics are used to enhance the power and potential of today's communication and security systems, providing a secure alternative to the conventional encryption methods that will resist also quantum computer attacks. While Quantum Teleportation and Quantum Computation mainly utilize entanglement between two or more particles, Quantum cryptography can also be performed even with single quantum particles.

Single photon communication

Single photon communication allows an ideally secure generation and distribution of a cryptographic key between two distant parties. Quantum Key Distribution (QKD) in fact uses the fundamental laws of quantum mechanics that describe the transmission of quantum states of the light [5–7]. The distribution of the cryptographic key must be secure against the attack of a third party that tries to acquire information with an eavesdropping technique. The vital advantage quantum mechanics provides lies in the impossibility that an eavesdropper (Eve) can intercept the secret key, made up of individual quanta, without revealing her presence to Alice, the sender and Bob, the receiver, since such interception unavoidably alters and destroys the quantum state of the photon. An attack may be made by Eve who secretly attempts to determine the key intercepting the travelling photons from Alice to Bob. By performing a sequence of measurements on these quanta, Alice and Bob determine the key they will use to encrypt their message. This aspect derives directly from the laws of quantum mechanics and it is also known as "No-cloning theorem", which states that it is not possible to duplicate a generic single quantum state without measuring it, thus without perturbing it in an irreversible way. The measurement on a quantum state in fact has not the meaning of revealing information coded in the quantum state, as it happens for the classical case. The sender Alice builds the cryptographic key with a sequence of single quanta prepared in different complementary quantum states. The easiest way is the use of the polarization states associated to single photons. The sender Alice randomly prepares the state of a photon and sends it to the receiver Bob. He then establishes the cryptographic key by independently performing a sequence of random polarization measurements on the photons. Both Alice and Bob will independently generate two random sequences of "0"s and "1"s corresponding to the different outcomes of their polarization measurements. The cross-correlation of the random sequences and the protocol chosen by Alice and Bob will generate the quantum key. The two major protocols are those by Bennett and Brassard [8] and Bennett [9] (see [10] for a review). The application of space and astronomical technologies to Quantum Communication will make possible the realization of a future quantum cryptography-based network of Satellites and ground stations which will guarantee a completely secure, global, long-range communication system: a novel field which will disclose entirely new ways of exchanging information between distant observers.

2. THE Q-SPACE PROJECT

The theoretical properties of quantum information and the feasibility of quantum communication in practical situations have been already elucidated by many experiments carried out in laboratories on the ground.

Our experiment points out the fundamental advantages to be obtained by using a Space system, advantages that could lead to a deeper understanding of the theory and to novel utilisation. The aim of this experiment is to realize the first quantum communication from satellite to ground with single photons. This link is realized with the Matera Laser Ranging Observatory (MLRO) in conjunction with an existing retroreflecting satellite such as Lageos [11]. A laser beam, collimated by MLRO, will illuminate the retroreflectors onboard Lageos and simulate the transmission of single photons from satellite to ground. After the retroreflection, the detection and characterization of the quantum state associated to each photon through MLRO itself will simulate the QKD process. In this experiment we simulate Alice, the sender, located on the satellite and Bob, the receiver, in the ground station.

2.1 Description of the experimental setup

2.1.1 Properties of MLRO laser ranging system.
The laser ranging station is equipped with a 1.5 meter mirror telescope with astronomical quality optics, Cassegrain configuration f/212, long Coudé. The beam divergence is "diffraction limited" and can be tuned in a continuous way from around 1" to 20". The alt-azimuth mount is capable of a tracking velocity of 20 deg/sec in azimuth and 5 deg/sec in elevation, with a tracking accuracy of 1 arcsec RMS. The MLRO laser is an active hybrid ND:YAG configured to emit 40 ps fast pulses in the wavelength of 532 nm with an impulse of 100 mJ/pulse in the monochromatic setup. The averaged estimated number of photons per pulse is $\sim 5 \times 10^{17}$. The clock of the laser system is 10 Hz, with a range timing accuracy better than 2 ns, which is too slow to realize a quantum key with single photons from the satellite. This immediately requires, that the pulse rate of the laser is increased by many orders of magnitude. At 10 Hz pulse rate, it would require 30 days to capture 1 single event for the best case.

2.1.2 Properties of the retroreflecting targets.
Here the ground based station MLRO acts as photon source for the orbiting Alice. To simulate Alice on the satellite, we will use mainly the Laser GEOdynamics Satellite 1 **LA-GEOS 1** (ASI/NASA), a spherical-shape satellite for geodesy equipped with corner-cube retroreflectors, diameter 60 cm, with perigee height of 5900 km.

The satellite is equipped with 426 retroreflectors built in such a way to send most of the signal back to the same direction of the incident light. Each retroreflector is essentially a cylindrical piece of glass having a 3.8 cm in diameter and a 3-surface reflecting corner.

2.1.3   The quantum transceiver.

We use the laser ranging facility of MLRO to center the satellite with high precision and its optical transmission line for our optical transmission/receiver device, the Quantum transceiver (QT).

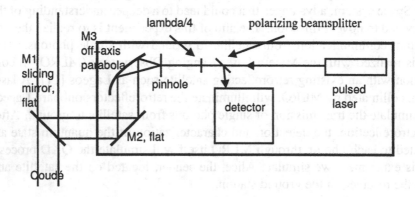

Figure 1.   Schematic of the Quantum transceiver.

QT is equipped with a weaker, but higher repetition rate laser, able to realize a quantum key distribution. The laser is a Nd:YAG laser with a passive Q-switching and integrated second harmonic generator, centered at 532 nm, with repetition rate of 17 kHz, which emits about 10^{12} photons per shot. The transmission and receiving line have the same optical paths, and use the polarization of light to discriminate the outgoing and incoming signals. The transmission is realized by opportunely polarized collimated pulses, focalized onto a pinhole, and expanded to the exit pupil diameter of the MLRO telescope by the off-axis parabola. After the retroreflection from the satellite, the beam passes again through a pinhole is recollimated with a lens, and directed by the polarizing beamsplitter to the detector, a Si-APD photon counter. To reduce the noise due to the light background (atmosphere, celestial sources, environmental, etc.) we choose a very small field of view, centered on the satellite position, with radius 1.4 to 2.2 arcsec. An additional improvement is given by the insertion of a narrowband filter with 0.15 nm of FWHM in the optical path to the detector and the time-tagging, obtained by setting up a set of time-windows, where to look for a positive detection by calculating the time of flight of laser pulse. This would allow us to label each possible photon reflected from the satellite, even if embedded in the background light, with a certain, finite, probability of success.

3. CATCHING THE PHOTON

3.1 Orbital fit and the link budget equation

The time of flight of the photons retroreflected by the satellite is varying in time, during the motion of the satellite itself. To determine exactly the set of coincidences for the time-tagging, we performed a polynomial fit obtaining only a few ns RMS difference between measured and fitted range points.

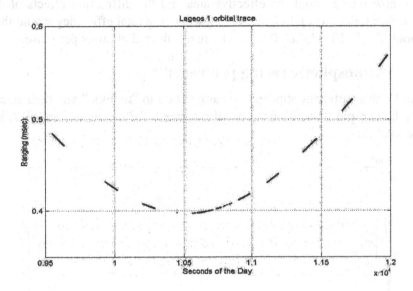

Figure 2. Lageos 1 : time of flight of a photon (from 40 to 58 ns) vs time.

A better determination of the time of flight can be obtained with Geodyn II (NASA/GSFC) program [5], with an actual error of 5-6 mm RMS in the position of the satellite. The post fit radial residuals for a good pass observed by MLRO is less than 1 cm RMS (often less than 5 mm RMS) for all targets. We now consider how to achieve single photon transmission from the satellite and single photon reception at the MLRO station by calculating the energy budget of the retroreflected light to choose the right energy to be transmitted. The returned energy is given by the link budget equation, which gives the number of reflected photons and then the number of photoelectrons, depending on different situations during the observation.

$$N_{phe} = \eta_q \, E_T \, \frac{\lambda}{hc} \, \eta_T \, G_T \, \sigma_{sat} \left(\frac{1}{4\pi R^2} \right)^2 A_T \, \eta_R \, T_A^2 \, T_c^2$$

here η_q is the detector quantum Efficiency, η_T the transmit path efficiency, η_R receive-path efficiency, A_T the receiving aperture area of the telescope, T_A the atmospheric transmission, T_c cloud transmission, E_T the laser energy pulse,

σ_{sat} the satellite backscattering cross section and G_T the transmit gain. A quite comprehensive description of the link budget calculation is given by Degnan [12]. The link budget is function of R^{-4}, R being the distance of the satellite from the earth (another way of expressing it, is by using the square of the beam divergence D). The main limiting factor of the efficiency is the size of the retroreflectors, which are only 3.8 cm in diameter. Furthermore, since the satellite is not stabilized, the retroreflectors must have some beam spreading. Taking now into account the effective area and the diffraction effects of the Lageos retroreflectors and of MLRO telescope, the total efficiency would then be about 10^{-16}–10^{-13}, i.e., 0.0001–0.1 retroreflected photons per pulse.

3.2   Atmospheric seeing problems

The Earth's turbulent atmosphere causes stars to "twinkle" and their intensity undergo rapid fluctuations. This also happens for the signal sent by Alice on the satellite.

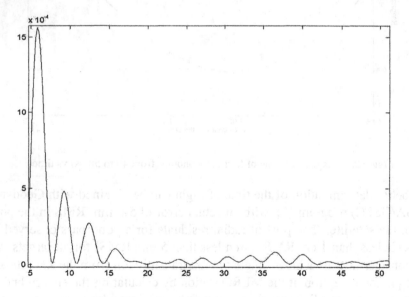

Figure 3.   Power Spectrum of Arcturus from 3 to 55 Hz.

The measured probability distributions for the arrival of the photons in time arise from a combination of the atmospheric fluctuations with the Poisson distribution of the photon counts. Previous studies on the analysis of atmospheric intensity scintillation of stars [13] show that the time distribution of the photon counts is quite complicated, which cannot easily fit neither with a Poissonian nor a Lognormal distribution. We characterized the effects due to the seeing, and the eventual loss in the transmission, by measuring the intensity of some test stars with our transceiver. In our power spectrum analysis of the signal of Arcturus, α Bootis, we find two peaks below 10 Hz, which can be ascribed to the guiding of the telescope. The effects of seeing, due to the atmospheric turbulence tilt of the wavefront, are in the peaks seen at higher frequencies.

3.3   Acquisition from ground targets

The calibration of our instrument was obtained by measuring the return signal from a selected ground target (corner cube), located at 42.25 m, i.e., with a time of flight of 150.92 ns for each photon. We obtained 100% of returns. In Fig. 4 we report the log of the number of the returned signal counts from 50000 pulses vs. the difference between the expected time of flight and the actual arrival time within 10 ns.

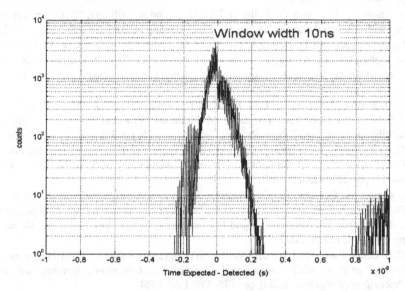

Figure 4. Logplot of the measured returns from a fixed target and difference between the time of flight and the detected time.

The main central peak contains almost all the returns and its width is caused by the non perfect regularity in the sequence of the laser pulses due to the Q-switching mechanism of the laser; other peak on the right is negligible, and can be caused by electronics bouncing. The data acquisition and analysis from orbiting satellites are now in progress.

4.   CONCLUSIONS

The first simple experiments made during the realization of the Matera-LAGEOS link, show that with nowadays technology the realization of a quantum cryptography link is feasible, without requiring much extra equipment, which can then serve as the basis for future large scale projects using dedicated space systems. This will be probably the first in the world experiment of such kind. This application represents the "quantum leap" that will transform a classical optical communication channel in a quantum channel, where it is possible to implement secure communication protocols based on Quantum Cryptography.

REFERENCES

[1] D. Bouwmeester, J.-W. Pan, M. Daniell, H. Weinfurter, M. Zukowski, and A. Zeilinger, "Reply: A posteriori teleportation," *Nature*, vol. 394, no. 6696, p. 841, 1998.

[2] G. Weihs, T. Jennewein, C. Simon, H. Weinfurter, and A. Zeilinger, "Violation of Bell's inequality under strict Einstein locality conditions," *Phys. Rev. Lett.*, vol. 81, no. 23, pp. 5039–5043, 7 Dec. 1998.

[3] T. Jennewein, C. Simon, G. Weihs, H. Weinfurter, and A. Zeilinger, "Quantum cryptography with entangled photons," *Phys. Rev. Lett.*, vol. 84, no. 20, pp. 4729–4732, May 2000.

[4] M. Aspelmeyer, H. R. Böhm, T. Gyatso, T. Jennewein, R. Kaltenbaek, M. Lindenthal, G. Molina-Terriza, A. Poppe, K. Resch, M. Taraba, R. Ursin, P. Walther, A. Zeilinger, "Long-distance free-space distribution of quantum entanglement," *Science*, vol. 301, no. 5633, pp. 621–623, Aug. 2003.

[5] W. T. Buttler, R. J. Hughes, P. G. Kwiat, S. K. Lamoreaux, G. G. Luther, G. L. Morgan, J. E. Nordholt, C. G. Peterson, and C. M. Simmons, "Practical free-space quantum key distribution over 1 km," *Phys. Rev. Lett.*, vol. 81, no. 15, pp. 3283–3286, Oct. 1998.

[6] W. T. Buttler, R. J. Hughes, P. G. Kwiat, G. G. Luther, G. L. Morgan, J. E. Nordholt, C. G. Peterson, and C. M. Simmons, "Free-space quantum-key distribution," *Phys. Rev. A*, vol. 57, pp. 2379–2382, 1998.

[7] J. Cerne, M. Grayson, D. C. Schmadel, G. S. Jenkins, H. D. Drew, R. Hughes, A. Dabkowski, J. S. Preston, and P.-J. Kung, "Infrared Hall effect in high-T_c superconductors: Evidence for non-Fermi-liquid Hall scattering," *Phys. Rev. Lett.*, vol. 84, no. 15, pp. 3418–3421, Apr. 2000.

[8] C. H. Bennett and G. Brassard, "Quantum cryptography: Public key distribution and coin tossing," in *Proc. IEEE International Conference on Computers, Systems, and Signal Processing*, (Bangalore, India), pp. 175–179, Dec. 1984.

[9] C. H. Bennett, "Quantum cryptography using any two nonorthogonal states," *Phys. Rev. Lett.*, vol. 68, no. 21, pp. 3121–3124, 25 May 1992 .

[10] D. Bouwmeester, A. Ekert, and A. Zeilinger, *The Physics of Quantum Information*, Springer, 2000.

[11] See for further information http://ilrs.gsfc.nasa.gov/

[12] J. J. Degnan, "Millimeter accuracy satellite laser ranging: A review," in *Contributions of Space Geodesy to Geodynamics: Technology* (D. E. Smith and D. L. Turcotte, eds.), AGU Geodynamics Series, vol. 25, pp. 133–162, 1993.

[13] D. Dravins, L. Lindegren, E. Mezey, and A. T. Young, "Atmospheric intensity scintillation of stars," *PASP*, vol. 109, pp. 173–207, 1997.

QUANTUM-AIDED CLASSICAL CRYPTOGRAPHY WITH A MOVING TARGET

Fabrizio Tamburini[1,3], Sante Andreoli[2], and Tommaso Occhipinti[1,3]

[1]*Dept. of Astronomy, University of Padova, vicolo dell'Osservatorio 2, I-35122 Padova, Italy.*
[2]*Magneti Marelli Holding S.p.A., Motorsport.*
[3]*Department of Information Engineering, University of Padova, via Gradenigo 6/B, I-35131, Padova, Italy.*

Abstract: We propose an encryption method obtained combining low-light optical communication, in the limit of quantum key distribution (QKD) techniques, and classical cryptography with pre-shared key. We present a toy-application to the telemetric data transmission Formula 1 racing.

Key words: optical communication; secure communication; cryptography; quantum cryptography.

1.   INTRODUCTION

The recent method proposed to create and distribute securely a quantum encryption key to send secure messages takes its vital inspiration from the basic laws of quantum mechanics. Quantum cryptography started with the studies by Bennett and Brassard in 1980s and by Bennett in 1992 [1, 2] as a new method for generating and distributing secure cryptographic keys using the properties of Quantum Mechanics. In contrast to existing methods of classical key distribution (CKD), quantum key distribution, QKD bases its security on the laws of nature. The impossibility of cloning or measuring a quantum state without inducing an irreversible collapse of its wavefunction ensures the build-up of a secure encryptographic key distribution between two parties. For a review see e.g. [3]. Similar experiments [4, 5] illustrated the feasibility of quantum encryption in practical situations. Free-space QKD was first realized [6, 7] over a small distance of 32 cm only with a point-to-point table top optical path, and recently improved in atmospheric transmission distances of 75 m [8] in daylight and 1 km [9] in nighttime over outdoor folded paths, where the quanta of light were sent to a mirror and back to the detector. A daylight quantum key distribution had been realized over a distance of 1.6 km by Buttler et al. [5]. Recently Aspelmeyer et al. realized a quantum key distribution over the Danube using entangled photons [10]. Several groups have also demonstrated QKD over multi-kilometer distances of optical fiber [11–17] and recently realized a version of the experiment "in the real world", in which Alice and Bob were connected with 1.45 km of optical fiber sharing entangled photons. The average raw key bit rate was found to be about 80 bits/s after error correction and privacy amplification. idQuantique, MagiQ technologies and NEC realized commercial applications of secure quantum key distribution [18–20]. MagiQ technologies guarantee, for example, a fast-generating quantum key rate of 10 keys per second. The field is now sufficiently mature to be commercially implemented and to be a tool in fundamental research beyond the foundations of quantum mechanics and basic physics [21,22].

2.   QKD TO UPDATE A "MOTHER KEY"

In this paper we suggest a simple procedure to aid the classical cryptographic methods with Quantum Cryptography, when the environmental conditions and/or the requirements of obtaining a long key in a short time strongly play against QKD. This procedure will increase, time-by-time, with the one-time-pad methods of Quantum Cryptography, the global security of the scheme. This method was studied to improve the security of bi-directional telemetry of race cars in view of possible, future, quantum computer attacks.

A classical cryptographic scheme can be reduced to three main quantities: m the message, k the key and c the code, with the corresponding random variables M, K and C that describe their statistical behaviours. The encoding $C = \text{Code}(M,K)$ and the decoding $M = \text{Dec}(C,K)$ are suitable deterministic processes which are described by a set of instructions τ that require a computational effort that depends both on the length of the cryptographic key and on the chosen protocol. Even if modern classical encryption protocols, based on the computational complexity of their encoding algorithms, still resist to the attacks made with nowadays technology, they will become vulnerable in the next future to the attacks of quantum computers, e.g. with Shor's algorithm (for a review, see [23]).

This problem will be avoided with a fast-generating QKD scheme that will change the key with a rate much faster than the computational time needed to break the code, without giving enough time to the cracker to get the encoded information. In environmental conditions with high bit error rate the application of this procedure will become more and more difficult giving more chances to the cracker to break the code.

We propose a simple thought experiment, in a noisy environment, as alternative to realize a secure fast encryption in real time using an ancillary key k' obtained via QKD, to update a classical pre-shared cryptographic key k with a simple change, such as bit shift or another more complex cyphertext method. k, the *mother cryptographic key*, can be previously securely loaded in a secure environment.

The updating process of k is a simple encryption of k itself with k', and the new key thus obtained, $k'' = \mathrm{Cod}(K,K')$, s.t. length$(k'') = $ length(k), is used to encrypt new messages. k' has a randomly variable length that depends on the efficiency the of QKD process: length$(k') \leq$ length(k).

This method will be less and less useful in the limit of the ideal case, i.e. when length$(k') = $ length(k). This satisfies the prescriptions of a perfect secure scheme for the encryption of k, and the space of obtainable new keys is the space of the messages $M = K$, $H = K'$ and the space of generated codes is given by the new key k''. In this limit we have the classical, safest, encryption procedure: the key has the same length as the message to be sent but it is also the case in which the methods of QKD are fully applicable.

In the worst case, if a run of the QKD updating process does not have success and $k' = \emptyset$, is the null set, the map Cod becomes the identity map and $k'' = k$, ensuring that the message will be in any case encrypted by the previous key.

The two encryption methods, classical and quantum, are combined together to increase the security of the transmission. The intrinsic weakness of the classical key distribution between two distant parties, can be aided by the impossibility of a third-party eavesdropping with QKD, while the failures due to the non-optimal environmental conditions which usually play against QKD in the building quickly a key in real time, are supported by previously selected cryptograms and CKD with a pre-shared key.

A possible application is the realization of a secure communication of telemetric data between two parties in relative motion. An example is a key exchange between an airplane and a ground station or between a race car and the Box. Real-time telemetric data transmission needs in fact an immediate and secure encoding process, which cannot be easily guaranteed at the moment using only QKD.

3. TOY-APPLICATION IN F1 RACING

Here we studied the possible application of this method to encrypt the telemetry of a race car during a Grand Prix. Alice is located at the Box, while Bob is driving the car. Alice and Bob initially share a secret classical cryptographic key k with length N exchanged before the race, and choose a classical protocol to realize a secure communication such as DES, 3DES or AES. The Quantum

Cryptographic Key k' obtained by exchanging single-photon pulses from the box line to the car during a passage close to the box has the purpose to give to the on-board computer only information to start cyphertext methods on the mother Key.

The immediate advantage is that with this method we can, in principle, realize quite a secure encryption only adopting a mother key 256 bit long, reducing the additional computational requirements to the onboard electronics.

To realize a single-photon QKD from the box to the car in a race track we also have to satisfy some precise requirements:

1. Strong weight constrains: no mechanically moving parts to realize polarization swapping no laser devices mounted into the car.

2. The usual procedure of Quantum key authentication utilizes the resources onboard the car and is convenient to embed them inside the telemetric data.

3. Data transmission must be classically encrypted even in the case of QKD failure.

4. The laser must works in non-visible light at 1550nm, and this procedure is safe for the driver: each pulse is in fact made with a very faint source, ideally from 0.1 to 1 photon per shot as required by QKD.

In the simplest case, the car is equipped with a set of passive detectors device realized with single-photon detectors, polarizers and fiber-injection optics as in the figure. The noise due to the environmental light is screened by a narrowband filter centred at the laser's wavelength of about 1550 nm. Each of the polarizers has different orientations, according with the chosen QKD protocol.

During each passage Alice tracks the car and tells Bob via radio to switch on the electronic control that will activate each of its detectors in a random sequence, which will realize Bob's polarization swapping. Alice sends a random sequence of polarized photon pulses. Alice and Bob will publicly announce their keys via telemetry and will decide whether to encrypt k.

With a commercial LiNbO3 modulator, Alice can produce ideally a random sequence of polarization swapping with a clock rate up to the GHz rate. Previous experiments showed a clock rate up to 1-MHz [5] for daylight QKD.

In the simplest case, the tracking could be realized by a fixed direction beam-expander, and the car, passing through the region illuminated by Alice's laser would capture some photons to realize the quantum key. With an angular beam width of 0.5° Bob obtains a laser beam expanded up to 20 cm at about 20 m of distance. This would need a fast and efficient QKD process. In fact, for a car travelling at 100 m/s at a distance of 20 m, the beam crossing time is the order the millisecond, which means that we would need 1 MHz of photon counting independently from its polarization state, to build a 256 bit key at each passage.

With a GHz modulator and a laser attenuation to 0.1–1 photons/pulse (as required by single-photon QKD) the total detection efficiency needed is 1/100. The emission of radiation by the car and the track at that specific wavelength can is considered almost constant within the tracking time.

A further step would be the application of adaptive optics to improve the pointing of Alice's source to Bob.

We could think to extend this procedure of cryptographic key updating also to the case in which the quantum communication channel is replaced with a faint source of polarized photons, even if low-light optical communication is in principle different from QKD.

4. CONCLUSIONS

We proposed a study of feasibility for a method of mixed quantum and classical cryptography with an application to race car telemetry encryption in real time. This method would guarantee the presence of a cryptographic key for a secure telemetry also when the quantum channel is affected by strong noise. We proposed to extend this procedure also when the distribution of the auxiliary key is realized with low-light optical communication.

ACKNOWLEDGMENTS

We would like to thank Prof. G. Cariolaro for his encouragement, comments and suggestions.

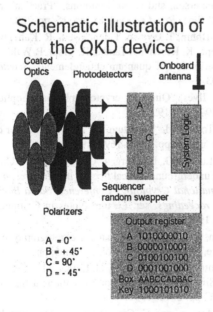

Figure 1. Scheme of the device onboard Bob's car.

REFERENCES

[1] C. H. Bennett and G. Brassard, "Quantum cryptography: Public key distribution and coin tossing," in *Proc. IEEE International Conference on Computers, Systems, and Signal Processing*, (Bangalore, India), pp. 175–179, Dec. 1984.

[2] C. H. Bennett, "Quantum cryptography using any two nonorthogonal states," *Phys. Rev. Lett.*, vol. 68, no. 21, pp. 3121–3124, 25 May 1992 .

[3] N. Gisin, G. Ribordy, W. Tittel, and H. Zbinden, "Quantum cryptography," *Rev. Mod. Phys.*, vol. 74, no. 1, pp. 145–195, Jan. 2002.

[4] R. J. Hughes, W. T. Buttler, P. G. Kwiat, S. K. Lamoreaux, G. L. Morgan, J. E. Nordholt, and C. Glen Peterson, "Practical quantum cryptography for secure free-space communications," Preprint: quant-ph/9905009, Available: http://arxiv.org/abs/quant-ph/9905009, 1999.

[5] W. T. Buttler, R. J. Hughes, S. K. Lamoreaux, G. L. Morgan, J. E. Nordholt, and C. Glen Peterson, "Daylight quantum key distribution over 1.6 km," Preprint: quant-ph/0001088, Available: http://arxiv.org/abs/quant-ph/0001088, 2000.

[6] C. H. Bennett, F. Bessette, G. Brassard, L. Salvail, J. A. Smolin, "Experimental quantum cryptography," in *Advances in Cryptology - EUROCRYPT '90, Workshop on the Theory and Application of of Cryptographic Techniques, Aarhus, Denmark, May 21-24, 1990, Proceedings*, (I. B. Damgård, ed.), ser. Lecture Notes in Computer Science, vol. 473, pp. 253–265, Springer, 1991.

[7] C.H. Bennett, F. Bessette, G. Brassard, L. Salvail and J. Smolin, "Experimental quantum cryptography," *J. Cryptology*, vol. 5, no. 1, pp. 3–28, 1992.

[8] B. C. Jacobs and J. D. Franson, "Quantum cryptography in free space," *Opt. Lett.*, vol. 21, no. 22, pp. 1854–1856, 1996.

[9] W. T. Buttler, R. J. Hughes, P. G. Kwiat, S. K. Lamoreaux, G. G. Luther, G. L. Morgan, J. E. Nordholt, C. G. Peterson, and C. M. Simmons, "Practical free-space quantum key distribution over 1 km," *Phys. Rev. Lett.*, vol. 81, no. 15, pp. 3283–3286, Oct. 1998.

[10] M. Aspelmeyer, H. R. Böhm, T. Gyatso, T. Jennewein, R. Kaltenbaek, M. Lindenthal, G. Molina-Terriza, A. Poppe, K. Resch, M. Taraba, R. Ursin, P. Walther, A. Zeilinger, "Long-distance free-space distribution of quantum entanglement," *Science*, vol. 301, no. 5633, pp. 621–623, Aug. 2003.

[11] J. D. Franson, and H. Ilves, "Quantum cryptography using optical fibers," *Appl. Opt.*, vol. 33, no. 14, pp. 2949–2954, 1994.

[12] C. Marand, and P. D. Townsend, "Quantum key distribution over distances as long as 30 km," *Opt. Lett.*, vol. 20, no. 16, pp. 1695–1697, 1995.

[13] R. J. Hughes, G. G. Luther, G. L. Morgan, C. G. Peterson, C. M. Simmons, "Quantum cryptography over underground optical fibers," in *Advances in Cryptology - CRYPTO '96, 16th Annual International Cryptology Conference, Santa Barbara, California, USA, August 18-22, 1996, Proceedings*, ser. Lecture Notes in Computer Science, vol. 1109, pp. 329–342, Springer, 1996.

[14] A. Muller, H. Zbinden, and N. Gisin, "Quantum cryptography over 23 km in installed under-lake telecom fibre," *Europhys. Lett.*, vol. 33, no. 5, pp. 335–339, 1996.

[15] R. J. Hughes, W. T. Buttler, P. G. Kwiat, G. G. Luther, G. L. Morgan, J. E. Nordholt, C. Glen Peterson, C. M. Simmons, "Secure communications using quantum cryptography," in *Proc. SPIE*, vol. 3076, pp. 2–11, 1997.

[16] R. J. Hughes, G. L. Morgan, and C. Glen Peterson, "Quantum key distribution over a 48 km optical fibre network," *J. Mod. Opt.*, vol. 47, no. 2/3, pp. 533–547, 2000.

[17] A. Poppe, A. Fedrizzi, T. Loruenser, O. Maurhardt, R. Ursin, H. R. Boehm, M. Peev, M. Suda, C. Kurtsiefer, H. Weinfurter, T. Jennewein, A. Zeilinger, "Practical quantum key distribution with polarization-entangled photons," Preprint: quant-ph/0404115, Available: http://arxiv.org/abs/quant-ph/0404115, 2004.

[18] idQuantique SA (Geneve, Switzerland), http://www.idquantique.com

[19] Magiq technologies (Sommerville, USA), http://www.magiqtech.com

[20] NEC Ltd. (Tokyo, Japan), http://www.nec.com

[21] F. Tamburini, C. Barbieri, S. Ortolani, and A. Bianchini, "Futuristic applications of quantum EPR states," in *Proc. Italian Astronomical Society*, vol. 74, no. 2, 2002.

[22] F. Tamburini and C. Barbieri, "Futuristic applications of quantum information and communication," in *Proc. Futuristic Space Technologies*, ASI workshop, 2002.

[23] D. Bouwmeester, A. Ekert, and A. Zeilinger (eds.), *The Physics of Quantum Information*, Springer, 2000.

II

CODING THEORY AND TECHNIQUES

CHANNEL CODING FOR OPTICAL COMMUNICATIONS

Invited Paper

Sergio Benedetto and Gabriella Bosco
Dipartimento di Elettronica, Politecnico di Torino, Corso Duca degli Abruzzi 24, 10129 Torino, Italy. E-mail: [lastname]@polito.it, Phone: +39-11-5644031, Fax: +39-11-5644099.

Abstract: Channel coding is a fundamental tool to improve performance in most of the digital transmission systems. By yielding the same performance with a significant saving in the transmitted power (the *coding gain*), channel coding helps increasing capacity in cellular systems, repeater spacing in terrestrial links, reducing the antenna size in deep-space communications, etc. It is now widely used also in optical communication, which, however, poses severe and peculiar challenges to the code design and implementation. The aim of this tutorial paper is to frame channel coding in optical systems by describing currently used coding schemes and strong future candidates promising higher coding gains. A rich annotated bibliography concludes the paper.

Key words: optical communication; channel coding; Reed-Solomon codes; turbo codes; low-density parity-check codes; block turbo codes; soft decoding; iterative decoding.

1.   INTRODUCTION

Channel coding has been known from more than half a century as a powerful mean to trade bandwidth with power using binary encoders, first, and then also to save power at no bandwidth expense with trellis-coded modulation (See [1], Chapters 10–12).

Central to the use of channel coding is the concept of coding gain, defined as the difference (normally expressed in decibels) between the signal-to-noise ratio per information bit needed to reach a desired bit error probability without and with coding, respectively. In Fig. 1 the bit error probability promised by information theory (owing to the famous Shannon Capacity Theorem) is plotted versus the minimum signal-to-noise ratio, expressed as the ratio between the energy per information bit E_b and the noise power spectral density

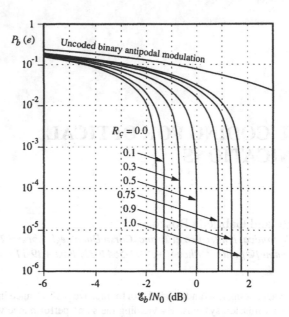

Figure 1. Bit error probability versus minimum signal-to-noise ratio for binary antipodal transmission over additive white Gaussian noise channel [1].

N_0. The channel is the additive white Gaussian noise (AWGN) channel. In order to evaluate the potential coding gain, also the uncoded curve referring to binary antipodal modulation is reported. Curves are labeled by the code rate R_c, defined as the ratio between the data rate at the input and the output of the channel encoder. Without bandwidth limitations (curve labeled by $R_c = 0$), coding gains up to 12 dB are theoretically possible at a bit error probability of 10^{-6}. Note that the coding gain increases for decreasing bit error probability because of the different slopes of coded versus uncoded curves.

During the past decades, channel coding has been used extensively in most digital transmission systems, from those requiring only error detection, to those needing very high coding gains, like deep-space links.

For quite a while, on the other hand, optical communication systems have ignored channel coding, until it became clear that it could be a powerful, yet inexpensive, tool to add margins against line impairments such as amplified-spontaneous emission (ASE) noise, channel cross talk, nonlinear pulse distortion, and fiber aging-induced losses. The use of channel coding in high-capacity wavelength-division multiplexed (WDM) fiber-optic links can then increase amplifier spacing, transmission distances and/or reduce optical power and the nonlinear effects that are tightly depending on it. Nowadays, channel coding is a standard practice in many optical communication links.

Turbo codes have been proposed for the first time in 1993 in a paper by Berrou *et al.* [2] that showed astonishing performance: a rate-1/2 turbo code together with a binary PSK (BPSK) modulation in an AWGN channel showed coding gains very close (0.5 dB) to the Shannon capacity limits using relatively simple component codes and large interleavers. The huge interest raised by turbo codes led researchers in the field to resurrect and improve the so-called low-density parity-check (LDPC) codes, proposed by Gallager [3]. The two classes of codes share a relatively simple, yet suboptimum, decoding algorithm that works recursively by performing local calculations and diffusing them back and forth to the other local atomized decoders until a high reliability on the information bits is achieved. Turbo and LDPC codes have revolutionized coding theory, and already found their way as standard in many important applications, like third generation cellular systems (W-CDMA), CCSDS telemetry channels, DVB-S2, and others. Their use in optical communication, as well as, for example, in magnetic recording, is still under heavy investigation and discussion.

This paper is tutorial in nature, and intends to describe the challenges of channel coding as applied to optical communication, the coding schemes that are presently used as de-facto standards, and the perspectives of employing turbo and LDPC codes as replacements with larger coding gains.

2. THE "STANDARD" CODING SCHEME AND ITS AVATARS

The use of channel coding in optical communication poses severe challenges to the designer. In fact, the code must guarantee at one time:

- Large coding gains (typically, greater than 6 dB) at very low bit error probabilities ($10^{-12} - 10^{-15}$)
- High code rates, with overheads typically lower than 25%
- Very high information rates, up to 40 Gbit/s.

The aforementioned set of requirements has driven the first proposals of channel codes for optical communication:

- Large coding gains at very low bit error probabilities require codes with large minimum distance, capable of avoiding the phenomenon known as "error floor", i.e., a significant reduction of the slope of the bit error probability curve at medium-low bit error probability values ($10^{-6} - 10^{-9}$). Algebraic block codes with large block sizes (to guarantee a large coding gain according to Shannon Theorem) are suitable candidates.

- High code rates also point to algebraic block codes, which in the range of code rates close to one behave better than convolutional codes.

- The very high information rates pose a severe complexity constraint, favoring hard decoding as opposed to soft decoding, which would require very fast ADC converters. Initially, the information rate requirement has

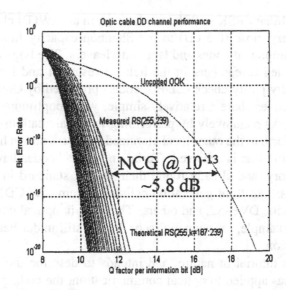

Figure 2.   Performance of the RS (255,239) code versus factor Q for uncoded and RS (255,239)-coded transmission [9].

been satisfied by demultiplexing the data flow into a suitable number of parallel flows at lower rate on which coding was applied, and successive multiplexing of the coded streams for optical transmission, with inverse operations at the receiver site. Typical information rates of the parallel flows was in the order of 1 Gbit/s, or slightly less. Recently, VLSI chips implementing co-decoders at information rates as high as 40 Gbit/s have been described [15].

It should then be clear why the first experiments of channel coding applied to optical communication were based on Reed-Solomon (RS) codes, which are non-binary, systematic linear cyclic codes, leading to the ITU G.975, then G.709 recommendation [17]. In it, the RS (255,239), characterized by a small overhead around 6.7% was suggested. This code, when hard-decoded with the standard algebraic decoding algorithm, yields a coding gain of 5.8 dB at a bit error probability of 10^{-13} (see Fig. 2, where the factor Q in dB is defined as $20 \log_{10} Q$). Other reasons for using RS codes with hard decoding are the availability of relatively low-complexity decoding algorithms, and well-understood and proved analytical models guaranteeing the performance at all bit error probabilities.

For high-capacity wavelength WDM long-haul systems using 40 Gbit/s channels increased coding gains are necessary, since higher bit-rate signals are more vulnerable to the fiber nonlinearity and dispersion and thus have less system margins. The coding gain can be slightly increased by moving some timid steps from hard to soft decoding [16]. To achieve significantly higher coding gains,

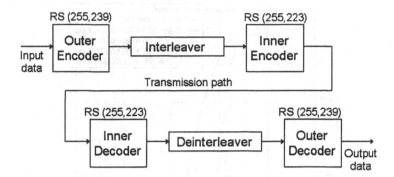

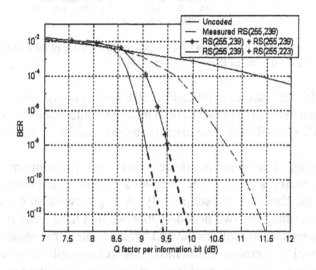

Figure 3. Block diagram of the concatenation of two RS encoders separated by an interleaver.

Figure 4. Performance of the concatenation of the code RS (255,239) with itself or with RS (255,223) versus factor Q [9].

the standard way is to add redundant bits to the code word, thus increasing the overhead and, consequently, the signal bandwidth and the related complexity of the electronic and optical devices. A "classical" solution suggested and tried for optical communication [19, 20] consists in the concatenation of two RS codes separated by an interleaver whose role is to spread the errors not corrected by the inner decoder over several outer encoder code words so as to increase the correction capability (see Fig. 3).

Concatenating two RS (255,239) codes leads to an overhead of 13.8% and yields a coding gain of roughly 7.4 dB. The concatenation of a RS (255,239)

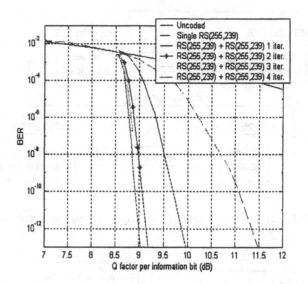

Figure 5. Performance of the hard iterative decoding algorithm applied to the concatenation of the code RS (255,239) with itself versus Q factor [9].

with a RS (255,223) has an overhead of 22% and a coding gain of 7.9 dB. In both cases, hard decoding applied to both inner and outer codes, an interleaving depth of 32 RS symbols and a bit error probability of 10^{-13} are assumed (see Fig. 4).

Hard decoding of the inner and outer codes is not the optimum (maximum-likelihood) decoding algorithm for concatenated codes, as Forney showed in his seminal PhD thesis on the subject [4]. A step forward in the right direction, still constraining the two decoders to work in a hard fashion, consists in iterating the decoding algorithms several times, hoping that residual errors at one iteration will be corrected in the next ones. An example is shown in Fig. 5, where the performance of the hard iterative decoding algorithm applied to the concatenation of the code RS (255,239) with itself is shown, with 1 to 4 iterations. The overhead is still the same as before, i.e. 13.8%, but the coding gain increases to 8.3 dB. There is no scope in increasing the number of iterations, as the gain saturates soon after.

3. THE IMPACT OF SOFT ITERATIVE DECODING

It is well known in the coding community that convolutional codes possess an inherent advantage over algebraic block codes since their maximum-likelihood (ML) decoding algorithm, the celebrated Viterbi algorithm [1], has almost the same complexity in its hard or soft versions. The same is far from true for algebraic block codes, for which ML soft decoding algorithms are order of magnitude more complex than the hard ones. Typically, soft versus hard

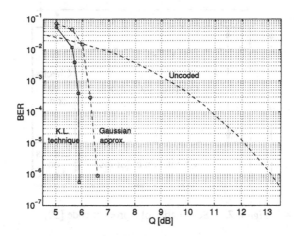

Figure 6. BER vs. Q factor after 5 iterations using a product code BCH(128,113)^2.

decoding offers an advantage in coding gain of roughly 2 dB. While soft decoding applied to the RS codes used so far in optical communications, although widely investigated in the recent literature [18], appears to be too complex at the data rates of interest, new forms of code concatenations with soft turbo-like iterative decoding have been proposed as the new frontier of channel coding for optical communication. In the following, we will describe the two most investigated (for optical applications) classes of codes endowed with iterative soft decoding algorithms, the so-called block turbo codes and the LDPC codes.

In both classes, the soft information from the communication channel to be used in the iterative decoding algorithms depends on the a-priori conditional probabilities, which in turn depend on channel noise statistics and receiver operations. If the communication channel can be properly modeled as an AWGN channel, the probability density function of the received signal is Gaussian and the log-likelihood ratio of the decision variable assumes a very simple form (proportional to the output of the receiver matched filter).

On the other hand, in long-haul amplified optical systems, the presence of a quadratic element (the photodetector) placed between the optical and the electric filters, leads to a strongly non-Gaussian noise at the output of the receiver [33]. Then, the exact expression of the a-priori conditional channel probabilities has to be found and used, in order to properly design an iterative decoding algorithm suitable to optical systems. In [33] it is shown that the modeling of the optical system as a Gaussian channel yields rather unreliable results when applied to the design of iterative soft decoding algorithms to be used in practical receivers. As an example, Fig. 6 shows the bit error probability (denoted by BER in the figure) at the output of an optical system as a function of the Q value. The solid line shows the results obtained evaluating

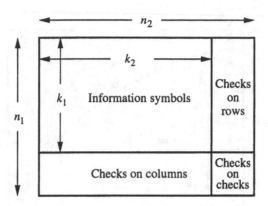

Figure 7.   Schematics of a product code [28].

the log-likelihood ratio using the Karhunen-Loeve (K.L.) technique described in [33], which takes into account all the propagation effects of the optical fiber, while the dashed line is referred to the results obtained evaluating the LLRs using the Gaussian approximation.

4.   TURBO PRODUCT CODES

The use of turbo product codes [28], a class of codes belonging to the wider class of serially concatenated codes with interleaver [5], characterized by two block codes concatenated with a classical row-column interleaver, has been at first proposed for optical communication systems in [29] .

Product codes are serially concatenated codes, composed by two systematic linear block codes with parameters (n_1, k_1) and (n_2, k_2), where n_i and k_i stand for codeword length and number of information bits respectively. The product code is obtained (see Fig. 7) by placing information bits in an array of rows and columns, coding the rows using code (n_1, k_1) and then coding the columns using code (n_2, k_2) . The parameters of the product code are $n = n_1 \times n_2$ and $k = k_1 \times k_2$. The code rate is given by the product of the two rates of the constituent codes, while the Hamming distance is equal to the sum of the Hamming distances of the constituent codes. It is thus possible to build very long block codes with large minimum Hamming distance by combining short codes with small minimum Hamming distance.

The sub-optimum, yet yielding a good performance/complexity trade-off, decoding procedure for such codes is based on iterating some sort of soft information on the reliability of information bits from one decoder to the other, until a reliable decision can be made.

The iterated soft information is, in its optimal form, the so-called *extrinsic* information, obtained by the a-posteriori probability of input (or output) bits to (from) each decoder. These quantities can be evaluated through the BCJR

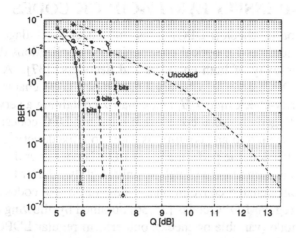

Figure 8. BER vs. Q factor after 5 iterations using a product code BCH(128,113,6)^2. The solid line shows the results obtained without quantization, while the dashed lines are referred to the results obtained using 2 (diamonds), 3 (stars), and 4 (circles) quantization bits.

algorithm [6], which is easily applicable to convolutional codes, but requires a larger effort for block codes, which do not possess a regular trellis configuration. In the case of block codes, it is customary to derive the soft information in a suboptimum way through a slight modification [28] of the Chase algorithm [7].

In [29–33] a product code composed by two (128,113) extended BCH codes was considered, which corresponds to a 28% overhead. The simulation results show a 7 dB coding gain of the product code with respect to the uncoded OOK at 10^{-6} (see Fig. 6), which corresponds to more than 2 dB gain with respect to the standard RS (255,239) code at the same bit error probability.

As already mentioned, optical communication systems require very high bit rates, which, in turn, make accurate A/D conversion a difficult issue. The effect of quantization on the performance has been evaluated in [29–33]. The results show that four bits of quantization yield almost ideal performance, whereas using only two bits corresponds to a 1.5 dB penalty. The performance of the TPC with 3 bit soft decision is superior by about 0.6 dB to that of TCP with 2 bit soft decision, which is superior by about 0.4 dB to that of the TPC with hard decision for 10^{-6} (see Fig. 8). As expected, better quantization improves the code performance, but A/D conversion is difficult at high speeds such as 10 Gbit/s. So, from a hardware implementation perspective, it can be concluded that 3 bits for the log-likelihood ratios representation is a reasonable performance/complexity trade-off [31]. The state-of-the-art result in the use of block turbo codes for optical communication is the experimental demonstration of a coding gain of 10.1 dB at 10^{-13} using 12.4 Gb/s block turbo code with 21% overhead and 3-bit soft decision [32].

5.   LOW-DENSITY PARITY-CHECK CODES

Low-density parity-check (LDPC) codes, introduced by Gallager in 1962 [3], are excellent candidates for optical communication applications due to the inherent low complexity of both encoders and decoders [37]. A low-density parity-check (LDPC) code is a binary linear block code characterized by a sparse parity parity-check matrix H, i.e., a matrix containing a very small number of ones. Gallager's original codes were *regular* LDPC codes: the number of ones in any row (and in any column) is equal to a fixed, very small number. In their original version, the performance of LDPC codes over the binary-input additive white Gaussian noise channel are only slightly inferior to that of parallel or serially concatenated convolutional codes (turbo-like codes as described in Section 2). Irregular LDPC codes [35,36] obtained by allowing different degrees for each node (variable or check), outperform regular LDPC codes and, for some values of the defining parameters, can outperform turbo codes.

In [34] it was shown that the encoding complexity of both regular and irregular LDPC codes is provably linear with the block size, without any restrictions on the parity-check matrix. Every LDPC code can be represented by a simple bipartite graph (a graph containing only two kind of nodes) composed by $(n-k)$ check nodes, corresponding to the code $(n-k)$ parity check equations (H rows), n variable nodes, corresponding to the n codeword bits (H columns) and a number E of edges connecting variable and check nodes (check node i and variable node j are connected if the bit h_{ij} of H is equal to one). LDPC codes can be decoded by various belief-propagation decoding methods such as one-step majority-logic decoding, bit-flipping, and iterative decoding based on the sum-product algorithm [8]. Although sum-product iterative decoding has been demonstrated to perform well in various types of channels, it is computationally intensive and it is not clear whether it is suitable for optical communications at data rate 40 Gb/s or not. However, the min-sum version of this algorithm (MSA), which is an approximation of the "a posteriori" probability decoding, requires only simple addition and "finding minimum" operations and, as such, is suitable for high-speed optical transmission.

In [37–39] the authors used finite geometric and combinatorial designs of LDPC codes because they require only simple encoders and decoders and because they assure large minimum codeword distances. In order to assess the performance of the proposed coding scheme, a very realistic simulation model was used, which takes into account in a natural way all major impairments in long-haul optical transmission such as amplified spontaneous emission (ASE) noise, pulse distortion due to fiber nonlinearities, chromatic dispersion, crosstalk effects, intersymbol-interference, etc. As noted in [33], this approach gives much better picture on the code performance than commonly used AWGN channel noise model [30].

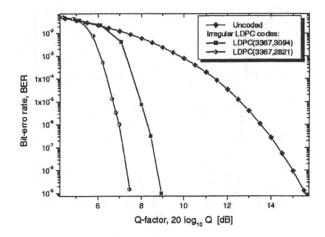

Figure 9. Irregular LDPC codes performance [39].

In [39] the authors compare the performance of two irregular LDPC codes (LDPC (3367, 3094) and LDPC (3367,2821)) with a standard RS (255,239) [23], showing that for BER of 10^{-9} the coding gains are 6.7 dB and 8.1 dB, 2–3.2 dB better than that of the RS (255,239) code (see Fig. 9). It is also shown that the irregular LDPC (3367,2821) gives comparable results with Turbo block code BCH $(128,113)^2$ [30] (of redundancy 28%) although it has smaller redundancy (19%). Much higher coding gains are expected for lower BERs.

In terms of quantization effects, LDPC codes seem to be more robust than TPC [40]. Comparing Fig. 10 with the previous Fig. 8 shows that for LDPC codes 2 bits of quantization yield already a very little loss.

A common problem in assessing the suitability of both turbo-like and LDPC codes to optical communication is the already mentioned phenomenon of error floor. Indeed, extrapolation of the curves obtained by simulation at bit error probabilities in the range $10^{-7} - 10^{-10}$, which are still obtainable by simulation, to lower values like 10^{-13} is a very dangerous and risky operation, since the slope of the curve can decrease very sensibly and lead to far too optimistic results.

Without resorting to simulation, an analytical estimate of the error floor behavior would be possible through the knowledge of the code minimum weight and the number of codewords with that weight. This, however, has been proved to be in general a problem with a non-polynomial complexity [13], and more viable ad-hoc solutions for LDPC codes are still an open research problem. More can be said for turbo-like codes and TPC [11]. A recent result [12] has shown that even the knowledge of the minimum distance does not provide reliable estimates of the error floor behavior, because of the suboptimality of the

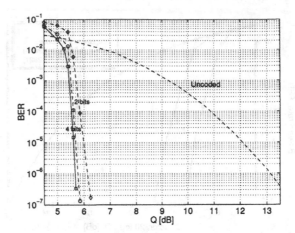

Figure 10.   BER vs. Q factor using an LDPC code LDPC(3276,2556), with a maximum number of iterations equal to 25. The solid line shows the results obtained without quantization, while the dashed lines are referred to the results obtained using 2 (diamonds) and 4 (circles) quantization bits.

iterative decoding algorithm. Indeed, while the minimum distance is strictly related to the asymptotic performance of maximum-likelihood decoding, the iterative turbo or message-passing algorithms can be thought of as maximum-likelihood algorithms working on an extended code, whose minimum distance can be significantly lower than the one of the original code.

The final word, then, is that the only way to claim the superiority of these new candidate channel codes for optical communication is to build fast decoders and let them work for a while, until bit error probabilities as low as 10^{-13} have been reliably estimated. By simulation, this task would require years even with state-of-the art optimized simulators.

A different approach consists in adding to turbo-like or LDPC codes an outer algebraic coding schemes such as BCH and RS, in order to improve the overall performance by lowering the error floor [14]. This approach is worth investigating also because the error floor of LDPC in the presence of burst of errors due to nonlinearities and signal dependent noise has not yet been assessed.

6.   HIGH-SPEED PARALLEL DECODER ARCHITECTURES

As already mentioned, very high-speed (up to 40 Gbit/s) RS hard decoders have been already implemented. The design of very high-speed iterative decoders, to reach decoding rates of the order of the Gbit/s, require highly parallel decoding architectures. The decentralized decoding structure for LDPC codes is highly suitable to parallel decoding, provided that the "collision" problem is avoided for parallel accesses to the memory banks. The same is true, although requiring a further effort of reducing the bi-directional sliding window soft-input soft-output (SISO) algorithm for turbo decoding to work on independent windows of data, for turbo-like codes. The collision problem in reading/writing from/into memory can be avoided by proper design of the parity-check matrix for LDPC codes (or of the interleaver for turbo-like codes) [41], or through a reworking of the addressing strategy that fits every LDPC or turbo-like encoder [46]. Examples of already implemented high-speed decoders can be found in [42–45]. A completely different approach to achieve very high decoding speeds consists in the analog implementation of the SISO algorithm using sub-threshold CMOS technology [48–51].

7. ANNOTATED BIBLIOGRAPHY

In this section, we list, for the reader who desires to delve deeper into the subject, a collection of papers categorized according to their subject. The first subsection contains basic papers on the subject of turbo and LDPC codes, as well as a pair of tutorial presentations on coding for optical communication. The second subsection deals with the use of RS codes for optical communication, and a recent paper of soft decoding for RS codes. The third one enumerates some recent papers reporting experimental results on optical communication systems using RS channel codes. The fourth subsection refers to block turbo codes, in general and applied to optical communication. The fifth one deals with LDPC codes, mainly applied to optical systems. In the sixth subsection we list decoder architectures designed to achieve a high throughput. Finally, in the last subsection, a few papers presenting different approaches in terms of coding schemes are inserted.

REFERENCES
Fundamental and tutorial papers

[1] S. Benedetto and E. Biglieri, *Principles of digital transmission with wireless applications*, New York: Kluwer Academic Publishers, 1999.

[2] C. Berrou and A. Glavieux, "Near optimum error correcting coding and decoding: Turbo-codes," *IEEE Trans. Commun.*, vol. 44, pp. 1261–1271, Oct. 1996.

[3] R. Gallager, "Low-density parity-check codes," *IEEE Trans. Inform. Theory*, vol. 8, no. 1, pp. 21–28, Jan. 1962.

[4] G. D. Forney, Jr., *Concatenated Codes*, MIT Press, Cambridge, Mass., 1966.

[5] S. Benedetto et al., "Serial concatenation of interleaved codes: performance analysis, design, and iterative decoding," *IEEE Trans. Inform. Theory*, vol. 44, no. 3, pp. 909–926, May 1998.

[6] S. Benedetto, D. Divsalar, G. Montorsi, and F. Pollara, "Soft-input soft-output modules for the construction and distributed iterative decoding of code networks," *European Transactions on Telecommunications*, invited paper, March-April 1998.

[7] D. Chase, "A class of algorithms for decoding block codes with channel measurement information," *IEEE Trans. Inform. Theory,* vol. 18, no. 1, pp.170–181, Jan. 1972.

[8] T. J. Richardson and R. L. Urbanke, "The capacity of low-density parity-check codes under message-passing decoding," *IEEE Trans. Inform. Theory,* vol. 47, no. 2, pp. 559–618, Feb. 2001.

[9] O. Ait-Sab, "Tutorial: Forward error correction techniques," in *Proc. OFC 2003,* Mar. 2003.

[10] F. Kerfoot, "Tutorial: forward error correction for optical transmission systems," in *Proc. OFC 2003,* Mar. 2003.

[11] R. Garello, F. Chiaraluce, P. Pierleoni, M. Scaloni, and S. Benedetto, "On error floor and free distance of turbo codes," in *Proc. of ICC 2001,* vol. 1 , pp. 45–49.

[12] R. Koetter and P. Vontobel, "Graph-covers and iterative decoding of finite length codes," in *Proc. 3rd Intl. Symp. on Turbo Codes,* Brest, France (Sept. 2003) Turbo conference, Brest 2003.

[13] A. Vardy, "The intractability of computing the minimum distance of a code," *IEEE Trans. Inform. Theory,* vol. 43, no. 26, pp. 1757–1766, Nov. 1997.

[14] A. Perotti, G. Montorsi, and S. Benedetto, "Performance analysis and optimization of concatenated block-turbo coding schemes," in *Proc. ICC 2004,* June 2004.

Reed-Solomon codes

[15] L. Song, M. Yu, and M. S. Shaffer, "10- and 40-Gb/s Forward Error Correction Devices for Optical Communications," *IEEE J. Solid-State Circuits,* vol. 37, no. 11, pp. 1565–1573, Nov. 2002.

[16] G. Bosco, G. Montorsi, and S. Benedetto, "A new algorithm for "hard" iterative decoding of concatenated codes," *IEEE Trans. Commun.,* vol. 51, no. 8, pp. 1229–1232, Aug. 2003.

[17] ITU-T Recommendation G.975: "Forward error correction for submarine systems," 1996.

[18] R. Koetter and A. Vardy, "Algebraic soft-decision decoding of Reed-Solomon codes," *IEEE Trans. Inform. Theory,* vol. 49, no. 11, pp. 2809–2825, Nov. 2003.

[19] H. Taga, H. Yamauchi, T. Inoue, and K. Goto, "Performance improvement of highly nonlinear long-distance optical fiber transmission system using novel high gain forward error correction code," in *Proc. OFC 2001,* paper TuF3, Mar. 2001.

[20] A. Agata, K. Tanaka, and N. Edagawa, "Study on the Optimum Reed-Solomon-Based FEC Codes for 40-Gb/s-Based Ultralong-Distance WDM Transmission," *J. Lightwave Technol.,* vol. 20, no. 12, pp. 2189–2195, Dec. 2002.

[21] J. Yan, M. Chen, S. Xie, and B. Zhou, "Performance evaluation of standard FEC in 40 Gbit/s systems with high PMD and prechirped CS-RZ modulation format," in *IEE Proc.-Optoelectron.,* vol. 151, no. 1, pp. 37–40, Feb. 2004.

Experimental applications of RS codes

[22] C.R. Davidson et al., "1800 Gb/s transmission on one hundred and eighty 10 Gb/s WDM channels over 7,000 km using full EDFA C-band," in *Proc. OFC 2000,* pp. PD25.1–PD25.3, Mar. 2000.

[23] J.-X. Cai et al., "2.4 Tb/s (120x20 Gb/s) transmission over transoceanic distance using optimum FEC overhead and 48% spectral efficiency," in *Proc. OFC 2001,* paper PD20, Mar. 2001.

[24] G. Vareille et al., "1.5 terabit/s submarine 4000 km system validation over a deployed line with industrial margins using 25 GHz channel spacing and NRZ format over NZDSF," in *Proc. OFC 2002*, paper WP5, Mar. 2002.

[25] X. Liu et al., "Enhanced FEC OSNR Gains in Dispersion-Uncompensated 10.7-Gb/s Duobinary Transmission Over 200-km SSMF," *IEEE Photon. Technol. Lett.*, vol. 15, no. 8, pp. 1162–1164, Aug. 2003.

[26] Y. Katayama and T. Yamane, "An experimental study of transmission distance limitation for the submarine cable system using C-band EDFA repeaters and a demonstration of 96WDM, 10 Gbit/s, over 12,000 km transmission having sufficient engineering margin," in *Proc. OFC 2003*, vol. 1, pp. 178–179, Mar. 2003.

[27] C. Rasmussen et al., "DWDM 40G transmission over trans-Pacific distance (10,000 km) using CSRZ-DPSK, enhanced FEC and all-Raman amplified 100 km ultrawave fiber spans," in *Proc. OFC 2003*, pp. PD18-1–PD18-3, Mar. 2003.

Block turbo codes

[28] R. Pyndiah, "Near-optimum decoding of product codes: block turbo codes," *IEEE Trans. Inform. Theory*, vol. 46, no. 8, pp. 1003–1010, Aug. 1998.

[29] O. Ait Sab and V. Lemaire, "Block turbo code performance for long-haul DWDM optical transmission systems," in *Proc. OFC 2000*, pp. ThS5-1–TuS5-4, Mar. 2000.

[30] O. Ait Sab, "FEC techniques in submarine transmission systems," in *Proc. OFC 2001*, pp. TuF1-1–TuF1-3, Mar. 2001.

[31] M. Akita et al., "Third generation FEC employing turbo product code for long-haul DWDM transmission systems," in Tech. Dig. Optical Fiber in *Proc. OFC 2002*, pp. 289–290, Mar. 2002.

[32] T. Mizuochi et al., "Experimental demonstration of net coding gain of 10.1 dB using 12.4 Gb/s block turbo code with 3-bit soft decision," in *Proc. OFC 2003*, pp. PD21-1–PD21-3, Mar. 2003.

[33] G. Bosco, G. Montorsi, and S. Benedetto, "Soft decoding in optical systems," *IEEE Trans. Commun.*, vol. 51, no. 8, pp. 1258–1265, Aug. 2003.

LDPC codes

[34] T. J. Richardson and R. L. Urbanke, "Efficient encoding of low-density parity-check codes," *IEEE Trans. Inform. Theory*, vol. 47, no. 2, pp. 638–656, Feb. 2001.

[35] T. J. Richardson, M. A. Shokrollahi, and R. L. Urbanke, "Design of capacity-approaching irregular low-density parity-check codes," *IEEE Trans. Inform. Theory*, vol. 47, no. 2, pp. 619–637, Feb. 2002.

[36] M. G Luby, M. Mitzenmacher, M. A. Shokrollahi, and D. A. Spielman, "Improved low-density parity-check codes using irregular graphs," *IEEE Trans. Inform. Theory*, vol. 47, no. 2, pp. 585–598, Feb. 2001.

[37] B. Vasic and I. B. Djordjevic, "Low-density parity check codes for long haul optical communications systems," *IEEE Photon. Technol. Lett.*, vol. 14, no. 8, pp. 1208–1210, Aug. 2002.

[38] I. B. Djordjevic, S. Sankaranarayanan, and B. V. Vasic, "Projective-Plane Iteratively Decodable Block Codes for WDM High-Speed Long-Haul Transmission Systems," *J. Lightwave Technol.*, vol. 22, no. 3, pp. 695–702, Mar. 2004.

[39] I. B. Djordjevic, B. V. Vasic, and S. Sankaranarayanan, "Regular and Irregular Low-Density Parity-Check Codes for Ultra-Long Haul High-Speed Optical Communications: Construction and Performance Analysis," in *Proc. OFC 2004*, paper WM4, Mar. 2004.

[40] G. Bosco and S. Benedetto, "Soft Decoding in Optical Systems: Turbo Product Codes vs. LDPC Codes," in *Proc. TIWDC 2004*, Pisa (Italy), pp. 79–86, Oct. 2004.

Decoders architectures

[41] M. M. Mansour and N. R. Shanbhag, "High-throughput LDPC decoders," *IEEE Trans. VLSI Syst.*, vol. 11, no. 6, pp. 976–996, Dec. 2003.

[42] A. Selvarathinam, E. Kim, and G. Choi, "Low-Density Parity-Check Decoder Architecture for High Throughput Optical Fiber Channels," in *Proc. ICCD'03*, pp. 520–525.

[43] H. Zhong and T. Zhang, "Design of VLSI Implementation-Oriented LDPC Codes," in *Proc. IEEE Semiannual Vehicular Technology Conference*, Oct. 2003

[44] B. Bougard, A. Giulietti, L. Van der Perre, and F. Catthoor, "A class of power efficient VLSI architectures for high speed turbo-decoding," in *Proc. GLOBECOM 2002*, vol. 1, pp. 549-553.

[45] R. Dobkin, M. Peleg, and R. Ginosar, "Parallel VLSI architectures and Parallel Interleaving Design for Low- Latency MAP Turbo Decoders," Technical Report CCIT-TR436.

[46] A. Tarable, G. Montorsi, and S. Benedetto, "Mapping interleaving laws to parallel Turbo decoder architectures," in *Proc. 3rd International Symposium on Turbo Codes and Related Topics*, pp. 153–156, Brest, France, Sep. 2003.

[47] M.J. Thul, F. Gilbert, and N. Wehn, "Optimized concurrent interleaving architecture for high-throughput turbo decoding," in *Proc. 9th Int. Conf. on Electronics, Circuits and Systems 2002*, vol. 3, pp. 1099–1102.

[48] J. Hagenauer et al., "Analog turbo-networks in VLSI: the next step in turbo decoding and equalization," in *Proc. 2nd Intl. Symp. on Turbo Codes*, Brest, France (Sept. 4-7, 2000).

[49] H.-A. Loeliger, F. Lustenberger, and M. Helfestein, F. Tarkoy"Probability propagation and decoding in analog VLSI," *IEEE Trans. Inform. Theory*, vol. 47, no. 2, pp. 837–843, Feb. 2001.

[50] V.C. Gaudet and P.G. Gulak, "A 13.3-Mb/s 0.35-/spl mu/m CMOS analog turbo decoder IC with a configurable interleaver," *IEEE J. Solid-State Circuits*, vol. 38, no. 11, pp. 2010–2015, Nov. 2003.

[51] C. Winstead, J. Dai, S. Yu, C. Myers, R. R. Harrison, and C. Schlegel, "CMOS analog MAP decoder for (8,4) Hamming code," *IEEE J. Solid-State Circuits*, vol. 39, no. 1, pp. 122–131, Jan. 2004.

Other codes

[52] H. Bulow, G. Thielecke, and F. Buchali, "Optical Trellis-Coded Modulation (oTCM)," in *Proc. OFC 2004*, paper WM5, Mar. 2004.

[53] Y. Katayama and T. Yamane, "Concatenation of interleaved binary/non-binary block codes fo rimproved forward error correction," in *Proc. OFC 2003*, vol. 1, pp. 391–393, Mar. 2003.

[54] P. Faraj, J. Leibrich, and W. Rosenkranz, "Coding gain of basic FEC block-codes in the presence of ASE noise," in *Proc. ICTON 2003*, pp. 80–83.

SOFT DECODING IN OPTICAL SYSTEMS: TURBO PRODUCT CODES VS. LDPC CODES

Gabriella Bosco and Sergio Benedetto
Dipartimento di Elettronica, Politecnico di Torino, Corso Duca degli Abruzzi 24, 10129 Torino, Italy. E-mail: [lastname]@polito.it, Phone: +39-011-5644036, Fax: +39-011-5644099.

Abstract: We consider the use of two classes of forward error correction (FEC) codes (concatenated codes with interleaver and LDPC codes) in optical communication systems. Soft iterative decoding is applied to both class of codes. In the simulations, the optimum log-likelihood ratio to be provided to the soft decoder in the optical channel environment is evaluated. Simulation results refer to practical turbo-product and LDPC codes, and encompass the effect of quantization on the log-likelihood ratio. The results show that LDPC codes give better results than turbo product codes. Moreover, they turn out to be more robust to quantization.

Key words: optical communication; low-density parity-check codes; product codes; soft-decision decoding; quantization.

1.   INTRODUCTION

Iterative decoding [1,2] is a powerful way of increasing the coding gain up to performance close to Shannon's theoretical limits. This can be obtained by using different concatenations of two (or more) generally simple constituent encoders with an interleaver in between. A sub-optimum decoding algorithm [2], whose arithmetic complexity is independent from the block size, has been presented. It iterates some sort of soft information from one decoder to the other and in a few iterations yields performance close to those obtainable by maximum-likelihood decoding.

Another class of block codes which exploits the advantages of iterative decoding is constituted by the so-called Low-Density Parity-Check (LDPC) codes [3]. In a recent paper [4] the authors show that the error performance and decoder hardware complexity can be greatly improved by using LDPC codes together with a soft iterative decoding process based on the *belief propagation* algorithm [5].

The soft information from the communication channel to be used in the iterative decoding algorithm depends on the a-priori conditional probabilities, which in turn depend on channel noise statistics and receiver operations. If the communication channel can be properly modeled as an additive white Gaussian noise (AWGN) channel, the probability density function (pdf) of the received signal is Gaussian and the log-likelihood ratio (LLR) of the decision variable assumes a very simple form (proportional to the output of the receiver matched filter).

On the other hand, in long-haul amplified optical systems, the presence of a quadratic element (the photo-detector) placed between the optical and the electric filters, leads to a strongly non-Gaussian noise at the output of the receiver [6]. It is clear that, in order to properly design an iterative decoding algorithm to be used in optical systems, the exact expression of the a-priori conditional channel probabilities has to be considered. In our simulations, we used the technique presented in [6] in order to properly evaluate such probabilities.

In Section 2, we describe the optical communication system analyzed in the paper. Sections 3 and 4 are devoted respectively to the description of the concatenated block codes and LDPC codes used in the simulations. Finally, in Section 5 some simulations results are presented. Since optical communication systems require very high bit rates, which, in turn, make accurate A/D conversion a difficult issue, we show also the effects of quantization on the codes performance [7,8].

2.   THE OPTICAL SYSTEM MODEL

A schematic representation of a digital optical communication system is shown in Fig. 1. The transmitter generates on-off keying pulses, which propagate along the optical fiber. At the receiver side, the signal passes through an optical pre-amplifier, followed by the optical filter $h_o(t)$, the photo-diode (modeled as an ideal square-law detector) and an electric filter $h_e(t)$ whose impulse response includes that of the photo-diode. The presence of optical amplification in the system makes the effects of shot noise and dark current negligible with respect to the Gaussian amplified spontaneous emission (ASE) noise introduced by the optical amplifiers [9]. The output of the electric filter is sampled at the optimum sampling instants $t_{opt} + nT$, where T is the symbol interval, to generate the soft sample sequence that represents the sufficient statistics to be used in subsequent processing.

When coding is used at the transmitter, one can either hard detect the sample sequence and provide the obtained digits to the decoder for hard decoding, or suitably process it to obtain the soft reliability measure to be provided to the decoder in the case of soft decoding.

In the case of binary codes and soft decoding, the optimum (maximum-likelihood or maximum-a-posteriori) decoder will be based on the knowledge of the *likelihood functions*, i.e., the pdfs of the signal at the electrical filter output conditioned on the transmission of a 0,1, and evaluated at the output y_n of the sampler [10].

In our simulations, we used the Karhunen-Loeve technique described in [6] in order to evaluate the exact log-likelihood functions to be passed to the soft iterative decoder.

3.   CONCATENATED BLOCK CODES

Stemming from the proposal of turbo codes [1], the application of concatenated block codes with interleaver in their serial form [2] known as *turbo product codes* to optical communication systems has been considered [7]. The sub-optimum, yet yielding a good performance/complexity trade-off, decoding procedure for such codes is based on iterating some sort of soft information from one decoder to the other, until a reliable decision on the information bits can be made.

The iterated soft information is, in its optimal form, the *extrinsic* information, obtained by the a-posteriori probability of input (or output) bits to (from) each decoder. These quantities can be evaluated through the BCJR algorithm (see [11] for a slightly more general description of it), which is easily applicable to convolutional codes, but requires a larger effort for block codes, which do not possess a regular trellis configuration. In the case of block codes, it is customary to derive the soft information through a slight modification [12] of the Chase algorithm [13]. The algorithm used in Section 5 is the one in [12]

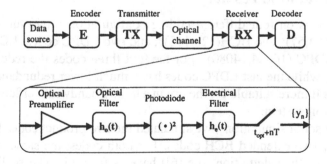

Figure 1.   Optical system schematics.

slightly modified in notations to accommodate a non Gaussian channel [6]. In one case, in Section 5 we have also used the optimum SISO algorithm.

4.   LOW-DENSITY PARITY-CHECK CODES

One major criticism concerning LDPC codes has been their apparent high encoding complexity. Whereas turbo codes can be encoded in a linear time, a straightforward encoder implementation for an LDPC code has a complexity that is quadratic in the block length [14]. In the design of our LDPC codes we used the technique at first proposed in [15], which makes the encoding complexity linear without the need of any pre-processing operation on the parity-check matrix. Moreover, in order to generate LDPC codes with good performance, we used the Progressive Edge Growth algorithm presented in [16].

The decoding algorithm used in Section 5 is the logarithmic version of the standard Sum-Product iterative decoding algorithm, described in [5]. Also in this case the decoding algorithm needs to be slightly modified in order to accommodate a non-Gaussian channel.

5.   APPLICATIONS AND SIMULATION RESULTS

5.1   Optical system description

The results reported in the following sections are obtained evaluating the bit-error rate (BER) of a 10 Gbit/s back-to-back optical system through direct error counting applied to the simulation model described in Section 2. The general scheme of an optical system employing FEC was shown in Fig.1. The encoded data are transmitted on the optical channel using an on-off keying (OOK) modulation technique with NRZ optical pulses (raised-cosine shaped with roll-off 0.8). The receiver is composed by a raised-cosine optical filter (roll-off 0.25) with bandwidth 20 GHz, followed by an ideal photo-diode and a 5-pole Bessel electric filter with bandwidth 8 GHz.

5.2   Simulation results

We will focus on four different coding configurations: two turbo product codes, BCH$(64,57)^2$ and BCH$(128,113)^2$, and two LDPC codes, LDPC$(3276, 2556)$ and LDPC $(16384,14080)$. For the first three codes the redundancy is around 28%, while the last LDPC codes has a much lower redundancy (16%), which makes it more suitable for the use in a highly bandwidth efficient optical communication system.

In the first set of simulations, we simulated the two product codes [12], each composed by two extended BCH codes. For both codes, the same simulation parameters (for their definition, see [6]) have been used. The scaling factor is $\beta = \{0.2, 0.3, 0.5, 0.7, 0.9, 1.0, 1.0, 1.0, 1.0, 1.0\}$. In the Chase algorithm,

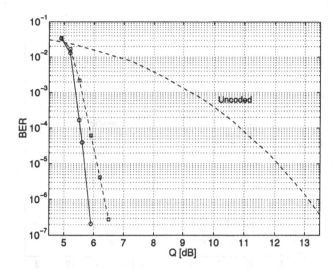

Figure 2. BER vs. Q factor after 5 iterations using a product code BCH(64,57)^2. Solid line: sub-optimum algorithm [12]. Dashed line: optimal SISO algorithm [11].

thirty-one test sequences have been used, and the parameter q has been chosen equal to 4. Five iterations have been used in the iterative algorithm, since the gain in terms of Q becomes negligible for additional iterations.

The BER as a function of the Q *factor* per information bit measured on the received electrical signal is reported in Fig. 2 for the BCH(64,57)^2. Two different decoding scheme have been used: the sub-optimum algorithm for near-ML decoding described in [12] (squares) and the optimal a-posteriori probability SISO algorithm described in [11] (circles). The loss due to the use of the approximate algorithm is of the order of 0.5 dB at 10^{-6}.

The results relative to the BCH(128,113)^2 are shown in Fig. 3. The solid line corresponds to the unquantized version of the decoding algorithm. To evaluate the effect of quantization on the performance, we show in Fig. 3 also the results obtained when using a finite, and low, number of bits to uniformly quantize the soft information at the output of the channel. The clipping threshold has been optimized for each number of quantization bits. The results show that four bits of quantization yield almost ideal performance, whereas using only two bits corresponds to a 1.5 dB penalty.

In the second set of simulations, we simulated two LDPC codes. The first one, LDPC(3276,2556), has the same rate as the BCH(128,113)^2 code. The maximum number of iterations in the decoding algorithm has been chosen equal to 25, so that the product between the code length and the number of iterations is the same as the one of the BCH code. The number of non-zero entries in the parity check matrix is 15120 (about 0.64% of the total number of bits in the matrix).

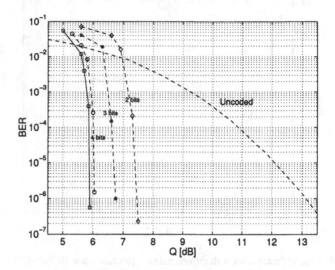

Figure 3.   BER vs. Q factor after 5 iterations using a product code BCH(128,113)^2. The solid line shows the results obtained without quantization, while the dashed lines are referred to the results obtained using 2 (diamonds), 3 (stars), and 4 (circles) quantization bits.

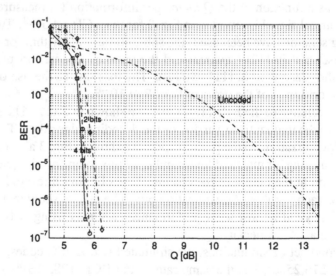

Figure 4.   BER vs. Q factor using an LDPC code LDPC(3276,2556), with a maximum number of iterations equal to 25. The solid line shows the results obtained without quantization, while the dashed lines are referred to the results obtained using 2 (diamonds) and 4 (circles) quantization bits.

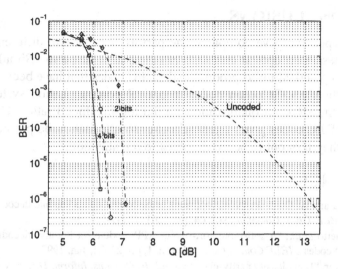

Figure 5.   BER vs. Q factor using an LDPC code LDPC(16384,14080), with a maximum number of iterations equal to 25. The solid line shows the results obtained without quantization, while the dashed lines are referred to the results obtained using 2 (diamonds) and 4 (circles) quantization bits.

The BER as a function of the Q *factor* per information bit measured on the received electrical signal is reported in Fig. 4. The solid line corresponds to the unquantized version of the decoding algorithm, while dashed lines have been obtained using a finite, and low, number of bits to uniformly quantize the soft information at the output of the channel. The clipping threshold has been optimized for each number of quantization bits. The results show that this LDPC code has better performance than the previously analyzed turbo product code (the gain is of the order of 0.5 dB at 10^{-5}). Moreover, it is more robust to quantization: using only two quantization bits, the loss is at 10^{-5} is less than 0.5 dB.

The second LDPC code, LDPC(16384,14080), has been chosen with a significantly higher rate than the other codes, in order to increase the bandwidth efficiency of the system. The maximum number of iterations in the decoding algorithm is again 25. The number of non-zero entries in the parity check matrix is about 0.14% of the total number of bits in the matrix.

The BER as a function of the Q *factor* per information bit measured on the received electrical signal is reported in Fig. 5. The solid line corresponds to the unquantized version of the decoding algorithm, while dashed lines have been obtained using quantization. The results show that this LDPC, which has the same codeword length as the BCH(128,113)^2, yields similar performance as the product code, but with a sensibly higher code rate.

6. CONCLUSIONS

In this paper, we have compared the performance of two different classes of FEC codes in an optical communication system scenario. Simulation results refer to practical turbo-product and LDPC codes that have been proposed to replace the Reed-Solomon codes in optical communication systems, and encompass the effect of quantization on the log-likelihood ratio. The results show that LDPC codes give in general better results than turbo product codes. Moreover, they turn out to be more robust to quantization.

REFERENCES

[1] C. Berrou and A. Glavieux, "Near optimum error correcting coding and decoding: Turbo-codes," *IEEE Trans. Commun.*, vol. 44, pp. 1261–1271, Oct. 1996.

[2] S. Benedetto *et al.*, "A soft-input soft-output APP module for iterative decoding of concatenated codes", *IEEE Comm. Lett.*, vol. 1, n. 1, pp. 22–24, Jan. 1997.

[3] R. Gallager, "Low-density parity-check codes," *IEEE Trans. Inform. Theory*, vol. 8 , no. 1, pp. 21–28, Jan. 1962.

[4] B. Vasic, I.B. Djordjevic, and R.K. Kostuk, "Low-density parity check codes and iterative decoding for long-haul optical communication systems," *J. Lightwave Technol.*, vol. 21, no. 2, pp. 438–446, Feb. 2003.

[5] T. J. Richardson and R. L. Urbanke, "The capacity of low-density parity-check codes under message-passing decoding," *IEEE Trans. Inform. Theory*, vol. 47, no. 2, pp. 559–618, Feb. 2001.

[6] G. Bosco, G. Montorsi, and S. Benedetto, "Soft decoding in optical systems", *IEEE Trans. Commun.*, vol. 51, no. 8, pp. 1258–1265, Aug. 2003.

[7] O. Ait Sab and V. Lemaire, "Block turbo code performance for long-haul DWDM optical transmission systems", in *Proc. OFC 2000*, pp. ThS5-1–TuS5-4, Mar. 2000.

[8] F. Zarkeshvari and A. H. Banihashemi, "On implementation of min-sum algorithm for decoding low-density parity-check (LDPC) codes," in *Proc. GLOBECOM '02*, vol. 2, pp. 1349–1353, 17–21 Nov. 2002.

[9] L. Kazovsky, S. Benedetto, and A. Willner, *Optical fiber communication systems*, London: Artech House, 1996.

[10] S. Benedetto and E. Biglieri, *Principles of digital transmission: with wireless applications*, New York: Kluwer Academic Publishers, 1999.

[11] S. Benedetto, D. Divsalar, G. Montorsi, and F. Pollara, "Soft-input soft-output modules for the construction and distributed iterative decoding of code networks", *European Transactions on Telecommunications*, invited paper, March-April 1998.

[12] R. Pyndiah, "Near-optimum decoding of product codes: block turbo codes," *IEEE Trans. Inform. Theory*, vol. 46, no. 8, pp. 1003–1010, Aug. 1998.

[13] D. Chase, "A class of algorithms for decoding block codes with channel measurement information," *IEEE Trans. Inform. Theory*, vol. 18, no. 1, pp.170–181, Jan. 1972.

[14] T. J. Richardson and R. L. Urbanke, "Efficient encoding of low-density parity-check codes," *IEEE Trans. Inform. Theory*, vol. 47, no. 2, pp. 638–656, Feb. 2001.

[15] Li Ping, W.K. Leung, and Nam Phamdo, "Low density parity check codes with semi-random parity check matrix," *Electronics Letters*, vol. 35, no. 1, pp. 38–39, Jan. 1999.

[16] Xiao-Yu Hu, E. Eleftheriou, and D.M. Arnold, "Progressive edge-growth Tanner graphs," in *Proc. GLOBECOM '01*, vol. 2, pp. 995–1001, 25–29 Nov. 2001.

ITERATIVE DECODING AND ERROR CODE CORRECTION METHOD IN HOLOGRAPHIC DATA STORAGE

Attila Sütő[1] and Emőke Lőrincz[2]
[1]Optilink Ltd, Bánhida 24, 1116 Budapest, Hungary, attila.suto@axelero.hu;
[2]Budapest University of Technology and Economics, Budafoki 8, 1111 Budapest, Hungary, lorincz@eik.bme.hu

Abstract: We propose a technique for data detection in two-dimensional page-access optical memory. The technique combines correlation decoding by the use of the Reed- Solomon algorithm with decision feedback to improve the symbol-error-rate performance in a system corrupted by intersymbol interference. By using the addition information of Reed-Solomon decoder, it is possible to increase efficiency of block decoding algorithm by correction the first order local errors in the received block code. We find experimentally that the feedback decision can improve data density by 11% during page recovery.

Key words: holographic data storage; intersymbol interference; channel modeling; decoding; error correction coding.

1. INTRODUCTION

We are developing a holographic memory system (HMS) based on polarization holography in thin-film photo-anisotropic polymer storage material [1]. A spatial light modulator (SLM) as two-dimensional data page using sparsely constant weight modulation coding feeds data into the system [2]. The information is stored in spatially filtered Fourier holograms. The coherent two-dimensional image is reconstructed by a pixelated detector (CCD) that converts the coherent images into electric signals. The original information can be recovered after decoding and error correction coding (ECC). In data storage systems the data density is the most important parameter. To increase data density higher signal to noise ratio (SNR) is necessary. More data processing time and redundancy that is appended to the source data can increase the SNR.

For optimum condition it is necessary to find a trade-off among data density, processing time and redundancy. A raw error ratio of 10^{-3} is suitable for ECC to achieve the acceptable 10^{-12} user error rate. This value can be reached from different SNR in function of decoding method and processing time consumption. The SNR and processing time can be optimized in a given HMS.

Conventional error-correction coding techniques [3,4] can be used to reduce the minimum signal-to-noise ratio that is required to achieve an acceptable symbol error rate and can therefore be used to increase the data density. Basically there are two approximations to solve the above-mentioned problem. First, the decoded data stream is feed into a Reed-Solomon ECC system [5] that can recover data through the noisy channel. This type of solution is effective when we have no information about the decoding defects. The second ECC method is the so-called Viterbi detection [5], which gives suitable result if the channel noise of the holographic system is known exactly. In fact, the real situation is that one part of noise comes from a well-described noise model; another part of noise may be estimated only during the decoding process.

In the following we present data organization of our HMS and the idea of an iterative decoding and feedback ECC process. We outline experimental system and the results of performed measurements with different ECC and decoding methods. We show effectiveness of our iterative decoding with ECC feedback to the achievable data density of the holographic system.

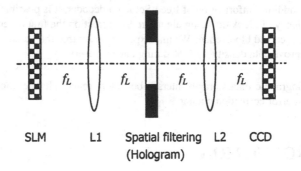

SLM	L1	Spatial filtering	L2	CCD
		(Hologram)		

Figure 1.   Schematic diagram of holographic memory system in the 4-f_L architecture.

2.   CHANNEL MODEL

The schematic of a 4-f_L (focal length), telecentric HMS is shown in Fig. 1. Each of the two lenses in the 4-f_L system performs a Fourier-transform (FT) operation, leading to the SLM being imaged onto the CCD. The aperture in the FT plane helps to minimize the blockage of useful signal and to maximize the storage density. The storage medium is placed priori to the first FT plane so that we can record a hologram. The system uses a grid of input-plane SLM pixels to represent binary 1's and 0's. We assume that the HMS is used in pixel-matched mode, wherein each SLM pixel is imaged onto one CCD pixel.

Several factors affect the pixel values output of the CCD. The optical aperture and the contrast ratio are main important factors that affect the CCD output value. Imaging imperfections of the optical system can be represented to the first order by the point-spread function (PSF). The PSF characterizes how much a point source spreads in consequence of the non-ideal imaging system.

The width of PSF is inversely proportional to the width of aperture in the frequency plane. If the input pattern in the object plane were a point source, the 2D field distribution in the output plane would correspond to the PSF. The PSF $h(x, y)$ for the system with a square aperture of width d would be

$$h(x, y) = \sin c\Big(\frac{xd}{\lambda f_L}, \frac{yd}{\lambda f_L}\Big). \qquad (1)$$

Here λ is the wavelength.

Interpixel cross talk, also called intersymbol interference (ISI), can arise owing to the finite optical aperture and the limited modulation transfer function of the system. ISI is systematic in nature and is therefore predictable for given system parameters, whereas optical scattering and electric noises are random. There are two models to study the effects of an aperture in the Fourier-transform plane of a 4-f imaging system. The first model is referred to as the convolution model. This model calculates the predicted spatial distribution at various planes in the imaging system by performing the appropriate convolutions of the input image distribution dictated by Fourier transform theory. The second model is referred to as the fast Fourier-transform (FFT) model. This model also calculates the predicted spatial distribution of the signal at the hologram plane (aperture plane) but differs from the convolution model in that a two-dimensional FFT routine is used rather than convolution integral in the computation. We used FFT model to study the above-mentioned holographic memory system, the model is useful for the inclusion of arbitrary data formats and apertures because it is not necessary to express these patterns in an analytic form, and because its execution requires relatively little computational time.

The two-dimensional FFT of the input plane (data page) was performed by using the Cooley-Tukey FFT algorithm. FFT's were performed on each of the rows and then on the columns to give the complex-valued distribution of the Fourier transform plane. The effect of the aperture in the Fourier transform plane was included by multiplying the Fourier-transform distribution by an aperture function, which consisted of pixels with amplitude one inside the aperture and amplitude zero outside the aperture diameter. Using this channel model it is possible to calculate the effect of spatial filtering and finite SLM contrast with different 2D data pages.

For a given aperture, SLM fill factor and noise level the Signal to Noise Ratio (SNR) is defined as

$$SNR = \frac{\mu_{on} - \mu_{off}}{\sqrt{\sigma^2_{on} - \sigma^2_{off}}}, \tag{2}$$

where μ_{on}, μ_{off} mean average value of the ON and OFF pixel distribution and σ^2_{on}, σ^2_{off} mean their variance.

It is desirable to know how the SNR as defined in Eq. 2 scales with symbol error rate (SER) for a holographic storage system where symbol means one code word created by the encoder. We present several methods in the following section to approve SER in function of SNR.

3.   SOURCE ENCODING AND DECODING

Input data are converted to binary matrix using Look-up table (LUT). The purpose of a modulation code is to constrain the digital data patterns that are recorded. Some patterns are more likely to be corrupted by the channel, while other patterns may be particularly helpful for a given detection scheme. In our system we found optimal size of block code and code patterns. We implemented 4 by 4 pixel block code using constant ON pixel ratio. Each pattern contains three ON pixels in order to maximize signal to noise ratio in the holographic channel. After read-out data follow the decoding process that identifies block codes with correlation decoding. Let the noisy received code word be a vector of length l, $\mathbf{x} = \mathbf{y} + \mathbf{n}$, where $\mathbf{y}$ is the transmitted code word and $\mathbf{n}$ is the noise vector imposed by the holographic recording-reconstruction process. An advantage of the constant weight modulation code (fixed ON pixels content per code word) is that we can implement Maximum Likelihood (ML) detection to find the most probable transmitted code word. Below we show how ML detection process helps us to choose correctly the minimum-distance valid code word. Using constant weight code, the decoding algorithm is simplified for the task of maximum searching. Eq. 3 defines the decoding process labelling with $K(\mathbf{x}, \mathbf{y})$ scalar function.

$$K(\mathbf{x}, \mathbf{y}) = \max_{\mathbf{y}_i}(\mathbf{x}\mathbf{y}_i^T)|_{\mathbf{x}}, \tag{3}$$

where $\mathbf{y}_i$ is the ith pattern in the Look-up table, $\mathbf{x}$ is vector representation of the detected pattern. The noise source generated errors can be eliminated using ECC. A Reed-Solomon code is arguably the most powerful tool in the family of block codes. The non-binary Reed-Solomon ECC is able to indicate the success of error correction. The ECC output is able to show defective patterns. The feedback of this information to correlation decoding is useful to find another pattern, which is the nearest to the previous pattern. The feedback increases the efficiency of the ECC. Fig. 2 shows the process of iterative decoding using the feedback ECC information.

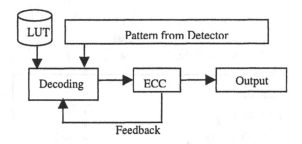

Figure 2. Data flow with feedback.

The ECC block error control code **C** consists of a set of **M** code words $\{y_0, y_1, y_2, \ldots, y_{M-1}\}$. Each code word is of the form $y = (y_0, y_1, \ldots, y_{l-1})$; if the individual coordinates take on values from the LUT. The encoding process consists of breaking up the data stream into blocks, and mapping these blocks onto code word in **C**. Mathematical representation of decoding and ECC can be written in the following way:

$$\text{ECC}(\text{DEC}(\underline{x}_0), \text{DEC}(\underline{x}_1), \ldots, \text{DEC}(\underline{x}_{M-1})) \rightarrow \{y_0, y_1, \ldots, y_{M-1}\}. \quad (4)$$

Where DEC function represents the decoding process formulated with Eq. 3. ECC function name represents Reed Solomon block decoder process. In general case this process can be implemented without any feedback decision. However, the iterative feedback decision increases the effectiveness and noise tolerance The feedback decision can be written mathematically in the following way:

$$\text{ECC}(\ldots \{\text{DEC}(\underline{x}_k), K^M(x, y)\}, \ldots) \rightarrow \{\ldots, \underline{y}_k^0, \ldots\}$$
$$\text{ECC}(\ldots \{\text{DEC}(\underline{x}_k), K^{M-1}(x, y)\}, \ldots) \rightarrow \{\ldots, \underline{y}_k^1, \ldots\}$$
$$\vdots \quad \vdots \quad \vdots$$
$$\text{ECC}(\ldots \{\text{DEC}(\underline{x}_k), K^{M-n}(x, y)\}, \ldots) \rightarrow \{\ldots, \underline{y}_k^n, \ldots\} \quad (5)$$

Denote $\{\text{DEC}(\underline{x}_k), K^{M-n}(x, y)\}$ expression the kth result of the decoding process with $K^{M-n}(x, y)$ as correlation values of the decoding process for the different code words in increasing order. The n value depends on the ECC feedback information, if RS algorithm indicates success during decoding then $n = 0$, else ECC decoding process is repeated while there is a successful ECC indication or limited by fixed iteration step. Of course data transfer rate will increase with the n value. The advantage of the above-mentioned algorithm is that the RS ECC with fixed parameter is more efficient than without the feedback decision method. It is important to find the trade-off between data transfer rate and tolerance of the noisy channel during optimization of system parameters.

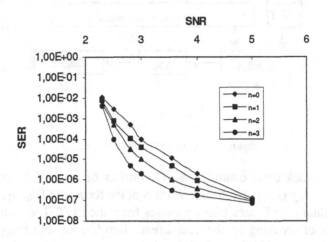

Figure 3.   SER in function of SNR and iteration steps.

4.   SIMULATION RESULTS

We show decision algorithms for decoding method with feedback and without feedback. By using the above-mentioned system model different decoding methods could be compared. The system SNR is estimated by using FFT model of the $4\text{-}f_L$ Fourier imaging system shown in Fig. 1. The output of the simulation is the SER. The data page consists of 320×240 pixels with 15 μm pixel sizes. The data page is divided into 4 by 4 pixels, which represents one code word. We used Fourier objective with a focal length of 4.4 mm. The laser wavelength is 532 nm. During simulation we used interleaved RS encoder/decoder with (60,52)(60,54) parameter.

Fig. 3 shows simulated SER in function of the SNR for different steps of iteration. The $n = 0$ is the result of correlation decoding and ECC without feedback. The result shows that there is a range in the SNR where the feedback decision is very efficient. In the SNR range of $2.5 \div 4$ the value of SER significantly improves in function of the SNR. It comes from the properties of constrained modulation codes. In the beginning of third section we mentioned that we use 4 by 4 pixels as code word. This block code contain only 8 bits so there are several unused code words in the LUT. If the received code word equals to an unused code word then it is easy to correct it by using feedback decision. However, the received code word is in the LUT then more iteration is required to get the right code word.

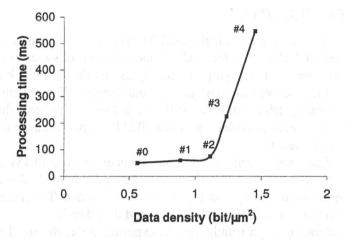

Figure 4. Data density in function of processing time consumption for target SER = 10^{-6}.

5. EXPERIMENTAL RESULTS

The feedback decision was implemented on Texas Instrument's TMS320C6201 DSP processor with a Co-Optic Coic5130A RS encoder/decoder and tested in our HMS.

We compared different methods of decoding and ECC performing experiments with the HMS and determining data density and processing time for a target SER of 10^{-6}. Fig. 4 shows how data density depends on the processing time when using different decoding methods. The #0 means the result of the global threshold method with ECC. The #1 is the result of the calculation using local threshold decoding method with ECC. Local threshold method means that the received code word is sorted in decreasing order and the bits associated with the largest NON received values (constant weight code) are set to 1, the remaining positions are set to 0. For #2 we used the correlation-decoding method (see Eq. 3) without any feedback from the ECC. #3 shows the effect of feedback ($n = 1$) ECC processing method. #4 presents the result with iterative ($n > 1$) feedback ECC.

Comparing different methods can be concluded that the feedback decision can increase the data density with about 11% in contrast to correlation decoding. Whereas the iterative decoding with feedback seems capable of providing as much as 18% density gain compared to simple feedback decision. In contrast to simple threshold method, block based modulation codes improve data density without significant processing time overhead.

6.   CONCLUSION

We have shown a simple channel model for the holographic data storage system. We used FFT algorithm for analysis and selection of system parameters for 4-f Fourier-transform imaging system. Sparse modulation block code was shown as an efficient way to construct 2-D data pages. We used small block length so a look-up table can be an effective solution. We have shown that the decoding algorithm provides maximum-likelihood performance in order to constant weight coding.

We have described a feedback decision detector for use in ISI-corrupted page-access Fourier holographic memories. The detection method is 2-D, employing correlation decoding and Reed-Solomon decoder. The incorporation of feedback improves the SER performance and data density.

We have shown through simulation and experiment that the novel decoding method can operate at a reasonable SER in a system. We found that a factor-of-2 improvement in data density can be readily achieved in contrast to simple threshold method with ECC.

REFERENCES

[1] E. Lőrincz, G. Szarvas, P. Koppa, F. Ujhelyi, G. Erdei, A. Sütő, P. Várhegyi, Sz. Sajti, Á. Kerekes, T. Ujvári, P. S. Ramanujam, "Polarization holographic data storage using azobenzene polyester as storage material," *Proc. of SPIE 4991 Organic Photonic Materials and Devices VI.*, J. G. Grote and T. Kaino eds., pp. 34-44.

[2] Brian M. King and Mark A. Neifeld, "Sparse modulation coding for increased capacity in volume holographic storage," *Appl. Opt.*, vol. 39, pp. 6681–6688, 2000.

[3] John F. Heanue, Korhan Gürkan, and Lambertus Hesselink, "Signal detection for optical memories with intersymbol interference," *Appl. Opt.*, vol. 35, pp. 2431–2438 ,1995.

[4] Mark A. Neifeld and Mark McDonald, "Error correction for increasing the usable capacity of photorefractive memories," *Optics Letters*, vol. 19, no. 18, 1994.

[5] Stephen B. Wicker, *Error Control Systems for Digital Communication and Storage*, chap. 8,12,16, Prentice Hall, New Jersey, 1995.

PERFORMANCE OF OPTICAL TIME-SPREAD CDMA/PPM WITH MULTIPLE ACCESS AND MULTIPATH INTERFERENCE

B. Zeidler, G. C. Papen, and L. Milstein*
Department of Electrical and Computer Engineering, University of California at San Diego, 9500 Gilman Drive, La Jolla, CA 92093-0407

Abstract: A novel optical modulation scheme employing ultrashort light pulses, modulated with Pulse Position Modulation (PPM) in conjunction with time-spread Code Division Multiple Access (CDMA) has recently been investigated in a two part paper by Kim, et.al. Mode-locked lasers producing sub-picosecond pulses every few nanoseconds are an excellent communication source, providing many terahertz of coherent bandwidth and a long time interval to spread out the pulse energy. Time-spread CDMA provides asynchronous multiple-access, while M-ary PPM provides enhanced throughput, affording up to several Tb/s of aggregate capacity.

In the previous work, Multiple Access Interference (MAI) was the only source of degradation considered, and the impact of propagation through a multipath channel was not addressed. Since PPM is inherently sensitive to multipath interference, any uncompensated stochastic phenomena producing multipath in the channel are likely to be the predominant sources of performance degradation. In this work, we develop a general analysis for the noncoherent detection of PPM with Gaussian MAI transmitted through a channel with two paths. As a specific example, we consider a single-mode, (chromatic) dispersion compensated fiber, where Polarization Mode Dispersion (PMD) is the only significant source of multipath. To a first-order, this produces two distinct paths with a stochastic Differential Group Delay (DGD) that is approximately Maxwellian distributed.

Assuming a polarization insensitive receiver, the statistics of the intensity samples at the M possible pulse positions are derived. The sum of the M-1 conditional pairwise error probabilities forms a union bound for the symbol error probability. There are three cases of interest, which are bounded analytically. There is no practical upper bound for the third case, so an approximate solution is obtained by numerically inverting the characteristic function. Lastly, a bound on the total unconditional error probability is obtained.

Key words: pulse position modulation; code division multiaccess; time-spread; multipath interference; multiaccess interference; polarization-mode dispersion.

*This work was partially supported by NSF grants NSF-9813721 andNSF-0123405.

1.   SYSTEM DESCRIPTION

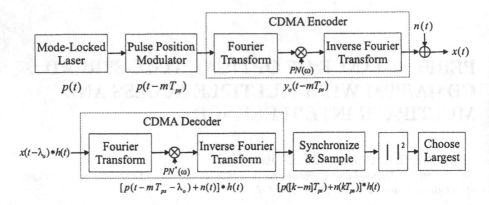

Figure 1.   Block Diagram

Throughout this paper, t denotes a real variable with units of seconds, x is a *dimensionless* real variable, and the letters $[i..n]$ represent dimensionless integers. We will define several dimensionless variables and functions normalized by the ultrashort pulse duration, which is the smallest timescale of interest.

We begin by summarizing the modulation and demodulation process for the desired user $i = 0$, illustrated above in Fig. 1. The original work in [1] contains detailed definitions and analysis of each signal:

1. Every T_s seconds, a mode-locked laser produces an ultrashort pulse $p(t) = \sqrt{P_o} p_u\left(\frac{t}{\tau}\right)$ which has

   - Peak power P_o,
   - Full Width Half Maximum (FWHM) duration τ seconds.
   - Normalized, dimensionless pulse shape function $p_u(x)$, with
        Unity peak value and FWHM duration,
        Autocorrelation function $R_p(x)$.

2. $p(t)$ is first pulse-position modulated to the mth of M timeslots that are separated by T_{ps} seconds, producing $p(t - mT_{ps})$.

3. $p(t - mT_{ps})$ is then time-spread by multiplying by a unique signature sequence sequence in the spectral domain.

   - Compared to $p(t)$, the peak power of $y_o(t)$ is reduced by the processing gain, N_{eff}, and the duration increases by approximately the same factor.

4. J-1 other asynchronous users produce MAI. For large J and N_{eff}, it can be shown that the MAI is approximately a stationary Gaussian process, $n(t)$ with zero mean and

- Variance $\sigma^2 \approx \frac{P_o}{N_{\text{eff}}} \frac{(J-1)}{\Omega T_s}$,
- Autocorrelation $R_n(t_o) \approx R_p\left(\frac{t_o}{\tau}\right)$.

5. The aggregate signal $y_o(t - mT_{ps} - \lambda_o) + n(t)$ is transmitted through a 2-path channel, $h(t)$.

6. The decoding spectral filter recovers the shifted ultrashort pulse $p(t - mT_{ps} - \lambda_o)$ from the time-spread pulse $y_o(t - mT_{ps} - \lambda_o)$;

   - The MAI is uncorrelated and remains *statistically* identical.
   - λ_o is a random delay removed by proper synchronization.

7. The intensity of $[p(t - mT_{ps}) + n(t)] \star h(t)$ is sampled every T_{ps} seconds, producing $I_k = |[p([k - m]T_{ps}) + n(kT_{ps})] \star h(t)|^2$.

8. The timeslot $\tilde{m}$ with the largest intensity, $I_{\tilde{m}}$ is selected as the most likely pulse position. When $\tilde{m} \neq m$, a symbol error is made.

2.   CHANNEL MODEL

As a specific example, we consider a Single Mode Fiber (SMF) of length L km with chromatic dispersion compensation. Considering only the first-order group-delay from PMD with coefficient D_p (in ps/$\sqrt{\text{km}}$) , it has been shown [2] that the channel can be modeled as a two path channel with a Maxwellian distributed differential delay:

$$|\hat{h}(t)|^2 = \alpha\, \delta(t) + (1 - \alpha)\, \delta(t - \tau_d) \qquad (1)$$

where α is uniformly distributed in $(0,1)$ and τ_d is Maxwellian distributed with mean $E\{\tau_d\} = D_p\sqrt{L}$, variance $\sigma_d^2 = \frac{\pi}{8} E\{\tau_d\}^2 = \frac{D_p^2 L \pi}{8}$, and pdf [2,3]

$$f_{\tau_d}(\tau) = \sqrt{\frac{2}{\pi}} \left(\frac{\tau}{\sigma_d}\right)^2 \frac{1}{\sigma_d} \exp\left(\frac{-\tau^2}{2\sigma_d^2}\right). \qquad (2)$$

We introduce *normalized, dimensionless* parameters for the mean delay, $\gamma \equiv E\{\tau_d\}/\tau$, and delay variance, $\overline{\sigma_d^2} \equiv \frac{\pi}{8}\gamma^2$. For instance, $\tau = 100$ fs and $D_p = 0.1$ ps/km$^{-\frac{1}{2}}$ corresponds to $L = \gamma^2$ km.

2.1   Quantization of τ_d and system synchronization

Since the Maxwellian is a continuous distribution over the range $(0, \infty)$, it is convenient to quantize τ_d into an integer multiple of the pulse separation, T_{ps}, plus a remainder. That is, $\tau_d = (\Delta + \tau_\delta)T_{ps}$, where Δ is a nonnegative integer and τ_δ is continuous in the interval $(0,1)$. Note that Δ and τ_δ are both dimensionless, and when $E\{\tau_d\}$ is sufficiently greater than T_{ps}, the remainder τ_δ is approximately uniform.

It is assumed that the system is synchronized to the strongest path, (e.g., the maximum of α and $1 - \alpha$) so that (in the absence of noise), the largest intensity sample corresponds to the correct pulse location. Therefore, we redefine α as uniform over $(\frac{1}{2}, 1)$, and $1 - \alpha$ as uniform over $(0, \frac{1}{2})$ with $sign(\tau_d) = \pm 1$ equiprobably. It can be shown that this random sign has no effect on the symbol error rate, so we assume $\tau_d > 0$.

3.   INTENSITY SAMPLES

We now define the *normalized, dimensionless* pulse separation, $\beta \equiv \frac{T_{ps}}{\tau}$ so that $p(xT_{ps}) = \sqrt{P_o}p_u(x\beta)$ and the interference autocorrelation function is $R_n(xT_{ps}) = R_p(x\beta)$. We assume β is large enough that $R_p(\beta) \approx 0$ and $p_u(\beta) \approx 0$, ensuring that, in the absence of multipath, adjacent interference samples are independent and $p(t)$ contributes power to only one sample. For example, for a Gaussian pulseshape $p_u(x) = 2^{-4x^2}$, $R_p(x) = 2^{-2x^2}$ and if $\beta = 3$, then $R_p(\beta) = 2^{-18}$ and $p(\beta) = \sqrt{P_o}2^{-36}$.

Using linearity and the orthogonality of the polarization modes, it can be shown that $I_k = \left| \sqrt{P_o}p_u([k-m]\beta) + n(kT_{ps}) \right|^2 \star \left| \hat{h}(t) \right|^2$, which is the sum of the intensities from the two polarization modes. This is illustrated below in Fig. 2.

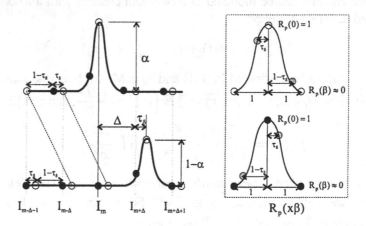

Figure 2.   Intensity Samples and Correlation Function with a delay of $(1 + \frac{1}{5})$ pulse positions.

Every sample I_k is the sum of two intensity samples (black and white) from the two arriving paths. The system is properly synchronized to the strongest path with the white samples aligned with the peak pulse value $\alpha\sqrt{P_o}$ at I_m, while the black samples from the weaker path are offset by $\Delta + \tau_\delta$ pulse positions. The correlation function $R_p(x\beta)$ is also shown in Fig. 2, with the gray circles illustrating the value of $\rho = R_p(x\beta)$. The normalized pulse separation

β is large enough that any two samples of the *same* color are uncorrelated, while *adjacent* samples with *opposite* color are *correlated*. Note that $\tau_\delta = 0$ (or 1) yields a degenerate, worst case scenario where the black and white samples are aligned and completely correlated.

Using $[N_1 \ldots N_4]$ to denote the four interference samples, the difference between the intensity at the correct pulse position, I_m, and at any other pulse position, I_k, is defined as:

$$D_k \equiv I_m - I_k \tag{3}$$
$$= \alpha|\sqrt{P_o} + N_1|^2 + (1-\alpha)|N_2|^2 - \alpha|N_3|^2 \tag{4}$$
$$- (1-\alpha)|\sqrt{P_o}p_u([k-m-\Delta-\tau_\delta]\beta) + N_4|^2$$

Thus, a symbol error is made if $D_k < 0$, since $I_m < I_k$, and the union bound is simply the sum of the pairwise error probabilities, $P_e = \sum_{k=1\ldots M}^{k\neq m} Pr\{D_k < 0\}$.

Note that indicies of the white samples are odd, and indicies of the black samples are even, which implies that only (N_1, N_4) or (N_2, N_3) can be correlated. Furthermore:

- The *only* black intensity samples that can be non-central are $k = m + \Delta$, $(\tau_\delta \to 0)$ and $k = m + \Delta + 1$, $(\tau_\delta \to 1)$.

- The *only* black intensity samples in I_k that can be correlated to the white sample in I_m are $k = m + \Delta$, $(\tau_\delta \to 0)$ and $k = m + \Delta + 1$, $(\tau_\delta \to 1)$.

- The *only* white intensity samples in I_k that can be correlated to the black sample in I_m are $k = m - \Delta$, $(\tau_\delta \to 0)$ and $k = m - \Delta - 1$, $(\tau_\delta \to 1)$.

Thus, for each value of $k \neq m$, D_k falls into one of three distinct cases of interest.

3.0.1   Case I: $k \neq \{m \pm \Delta, m \pm (\Delta + 1)\}$.   All of the interference samples are uncorrelated, and $p_u([k-m-\Delta-\tau_\delta]\beta) = 0$, therefore

$$D_k = \alpha|\sqrt{P_o} + N_1|^2 + (1-\alpha)|N_2|^2 - \alpha|N_3|^2 - (1-\alpha)|N_4|^2. \tag{5}$$

When $\alpha = 1$, there is no multipath and $D_k = |\sqrt{P_o} + N_1|^2 - |N_3|^2$, where $|\sqrt{P_o} + N_1|^2$ is a *non-central* χ^2 random variable (RV) with two degrees of freedom and $|N_2|^2$ is a *central* χ^2 RV with two degrees of freedom. This is identical to the error probability for noncoherent binary FSK and yields the lower bound:

$$P_{b,I}^{LB} = \frac{1}{2}\exp\left(-\frac{SNR}{2}\right). \tag{6}$$

When $\alpha = 1/2$, $I_m = \frac{1}{2}\left(|\sqrt{P_o} + N_1|^2 + |N_2|^2\right)$ is a non-central χ^2 RV with *four* degrees of freedom, and $I_k = \frac{1}{2}\left(|N_3|^2 + |N_4|^2\right)$ is a central χ^2 RV with *four* degrees of freedom as well. This yields the upper bound:

$$P_{b,I}^{UB} = \exp\left(-\frac{SNR}{2}\right)\left(\frac{1}{2} + \frac{1}{16}SNR\right). \tag{7}$$

3.0.2   Case II: $k = m - \Delta$ or $k = m - (\Delta + 1)$. D_k is identical to Case I, except that N_2 and N_3 can be correlated. For $k = m - \Delta$, $\rho = R_p(\tau_\delta\beta)$, and for $k = m - \Delta$, $\rho = R_p((1 - \tau_\delta)\beta)$. If $\rho = 0$, Case II reduces to Case I and with $\alpha = 1/2$ and we obtain the same upper bound $P_{b,II}^{UB} = P_{b,I}^{UB}$.

As the correlation increases for a given α, it *reduces* the error probability because N_3 partly cancels N_2. The lower bound is obtained for $\rho = 1$, (when $N_3 = N_2$) with $\alpha = 2/3$:

$$D_k = \frac{2}{3}|\sqrt{P_o} + N_1|^2 - \frac{1}{3}\left(|N_3|^2 + |N_4|^2\right). \tag{8}$$

Noting that $|\sqrt{P_o} + N_1|^2$ is a non-central χ^2 RV with *two* degrees of freedom, while $|N_3|^2 + |N_4|^2$ is χ^2 RV with *four* degrees of freedom, it can be shown that:

$$P_{b,II}^{LB} = \exp\left(-\frac{2}{3}SNR\right)\left(\frac{5}{9} + \frac{2}{27}SNR\right). \tag{9}$$

3.0.3   Case III: $k = m + \Delta$ or $k = m + \Delta + 1$.   We consider $k = m + \Delta$, noting that $k = m + \Delta + 1$ is identical except with $-\tau_\delta$ replaced by $1 - \tau_\delta$. $D_k = \alpha\left(|\sqrt{P_o} + N_1|^2 + |N_2|^2\right) - (1 - \alpha)\left(|N_3|^2 + |\sqrt{P_o}p_u(-\tau_\delta\beta) + N_4|^2\right)$, and N_1 and N_4 are correlated, with $\rho = R_p(\tau_\delta\beta)$. When $\alpha = 1$, there is no multipath and we obtain the same lower bound as in Case I, $P_{b,III}^{LB} = P_{b,I}^{LB}$.

When $\alpha = 1/2$ and $\tau_\delta = 0$, $p_u(\tau_\delta\beta)$ and ρ are both unity. Thus, $D_k = 1/2\left(|N_2|^2 - |N_3|^2\right)$ and the signal intensity is completely cancelled by the interference intensity, yielding $Pr\{D_k < 0\} = 1/2$. However, with $\alpha = 1/2$ and $\tau_\delta\beta \gg 0$ so that $\rho \approx 0$ and $p_u(\tau_\delta\beta) \approx 0$, then D_k is identical to the upper bound of Cases I and II. We conclude that the error rate is heavily influenced by $\tau_\delta\beta$, with no useful upper bound.

3.0.4   Case III Approximate Conditional Error.   Let $f_{D_k}(x|\alpha, \tau_\delta)$, $F_{D_k}(x|\alpha, \tau_\delta)$ and $\psi_{D_k}(\omega|\alpha, \tau_\delta)$ denote the conditional pdf, cdf and characteristic function of D_k, respectively. Although there appears to be no closed form solution for the condtional error probability, $P_{b,III}(\alpha, \tau_\delta) = F_{D_k|\alpha, \tau_\delta}(0)$, it can be found numerically by: Deriving $\psi_{D_k}(\omega|\alpha, \tau_\delta)$ analytically using the General Hermitian Quadratic form, sampling $\psi_{D_k}(\omega|\alpha, \tau_\delta)$ and computing the IFFT to find $f_{D_k}(x|\alpha, \tau_\delta)$, then integrating $f_{D_k}(x|\alpha, \tau_\delta)$ up to $x = 0$ to obtain $P_{b,III}(\alpha, \tau_\delta)$. This numerical technique has been experimentally validated for all 3 cases throughout the range of α and τ_δ using both Monte Carlo simulations and the analytical bounds.

3.0.5   Case III: Approximate Unconditional Error.   The unconditional error for case III is:

$$P_{b,III} = \int_0^1 \int_{1/2}^1 (P_{b,III}(\alpha, \tau_\delta) + P_{b,III}(\alpha, (1 - \tau_\delta)))\ d\alpha\ d\tau_\delta \qquad (10)$$

Since α and τ_δ are uniform RVs, the unconditional error probability can be approximated numerically by sampling α and τ_δ, solving the conditional error probability for each combination, and then computing the average. The results for a Gaussian pulse $p_u(x) = 2^{-4x^2}$ and $\beta = [1, 3, 10]$ are shown below in Fig. 3 (Left):

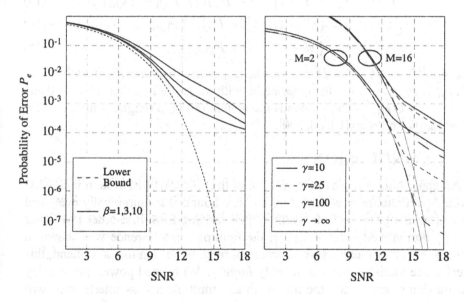

Figure 3.   Left: Case III unconditional error. Right: Upper bound vs. M and γ ($\beta = 3$).

Beyond $\beta > 3$, there is little performance improvement at high SNR. At 18dB, tripling the pulse separation (to $\beta = 9$) reduces the error rate by less than 15%.

3.1   Total error probability

The transmitted pulse position m and the integer multipath delay Δ govern how the 3 cases combine to form the total error probability. It can be shown that case III only affects the test statistic when $0 \leq m \pm \Delta \leq M$. Since the M

symbols are equally likely, this probability of interference is given by:

$$P_{int}(\sigma_d, M, T_{ps}) = \sum_{d=0}^{M} \left(\frac{M-d}{M} \right) [F_{\tau_d}((d+1)T_{ps}) - F_{\tau_d}(dT_{ps})] \quad (11)$$

$$\approx \int_0^{MT_{ps}} \left(\frac{M - x/T_{ps}}{M} \right) f_{\tau_d}(x)dx \quad (12)$$

where the integral approximation grows increasingly accurate as $T_{ps} \to 0$. For the Maxwellian distribution, we obtain $P_{int}(\xi) = \frac{2}{\xi\sqrt{\pi}}[e^{-\xi^2} - 1] + \text{erf}(\xi)$ where $\xi = \frac{MT_{ps}}{\sigma_d} = \frac{M\beta}{\gamma}\sqrt{\frac{8}{\pi}}$ is dimensionless.

It can be shown that the total error probability is a function of ξ, bounded by:

$$P_e(\xi) \leq (M-1)P_{b,I} + P_{int}(\xi)(P_{b,II} + P_{b,III}) \quad (13)$$

$$\leq (M-1)P_b^{UB} + P_{int}(\xi)P_{b,III} \quad (14)$$

$$\leq (M-1)P_b^{UB} + P_{b,III} \quad (15)$$

where $P_b^{UB} \equiv P_{b,I}^{UB} = P_{b,II}^{UB}$, and the last bound becomes tight with $\sigma_d \ll MT_{ps}$, as $P_{int} \to 1$. On the other hand, with $\sigma_d \gg MT_{ps}$, then $P_{int} \to 0$ and $P_e \approx (M-1)P_b^{UB}$. These bounds are plotted in Fig. 3 (Right) with $\beta = 3$ for $M = [2, 16]$ and $\gamma = [10, 50, 100]$.

4.   CONCLUSIONS

An upper bound on the performance of the CDMA/PPM system with MAI and a 2-path channel has been derived. The bound was numerically evaluated for a Gaussian pulse and Maxwellian DGD arising from first-order PMD. For a *small* normalized mean delay, γ, the multipath interference was shown to increase the error rate by several orders of magnitude. On the other hand, this interference vanishes for a sufficently *large* γ. With fixed power, mean delay and maximum error rate, the multipath and multiple access interference will impose a limit on the symbol size, M, and the number of users, J.

REFERENCES

[1] K. S. Kim, D. M. Marom, L. B. Milstein, and Y. Fainman, "Hybrid pulse position modulation/ultrashort light pulse code-division multiple-access systems - Part 1: Fundamental analysis," *IEEE Trans. Commun.*, vol. 50, pp. 2018–2031, Dec. 2002.

[2] M. Karlsson, "Probability density functions of the differential group delay in optical fiber comunications systems," *J. Lightwave Technol.*, vol. 19, pp. 324–331, Mar. 2001.

[3] M. Simon, *Probability Distributions Involving Gaussian Random Variables*, Kluwer Press, 2003.

PERFORMANCE ANALYSIS AND COMPARISON OF TRELLIS-CODED AND TURBO-CODED OPTICAL CDMA SYSTEMS

M. Kulkarni, P. Purohit, and N. Kannan
Department of Electronics & Communication Engineering
Delhi College of Engineering, Bawana Road, Delhi-110042, India

Abstract: In this paper, we have compared the performance of a Turbo coded and a Trellis coded optical code division multiple access (OCDMA) system. The comparison of the two codes has been accomplished by employing optical orthogonal codes (OOCs). The Bit Error Rate (BER) performances have been compared by varying the code weights of address codes employed by the system. We have considered the effects of optical multiple access interference (OMAI), thermal noise and avalanche photodiode (APD) detector noise. Analysis has been carried out for the system with and without double optical hard limiter (DHL). From the simulation results it is observed that a better and distinct comparison can be drawn between the performance of Trellis and Turbo coded systems, at lower code weights of optical orthogonal codes for a fixed number of users. The BER performance of the Turbo coded system is found to be better than the Trellis coded system for all code weights that have been considered for the simulation. Nevertheless, the Trellis coded OCDMA system is found to be better than the uncoded OCDMA system. We can thus conclude that Trellis coded OCDMA can be used in systems where decoding time has to be kept low, bandwidth is limited and high reliability is not a crucial factor as in local area networks. Also the system hardware is less complex in comparison to the Turbo coded system. Trellis coded OCDMA can be used without significant modification of the existing chipsets. Turbo coded OCDMA should however be employed in systems where high reliability is needed and bandwidth is not a limiting factor.

Key words: optical code division multiaccess; optical orthogonal codes; optical multiaccess interference; avalanche photodiodes; double hard limiters; trellis coded modulation; turbo codes.

1.   INTRODUCTION

Recently, all optical CDMA techniques have received a growing interest. Optical CDMA (OCDMA) allows multiple users to access the network asynchronously and simultaneously. From the practical view point the OCDMA network is gaining popularity, since it requires minimal optical signal processing and is virtually delay free. In OCDMA systems, the BER performance is degraded by the OMAI, which comes from all the other active users. This in turn ultimately limits the number of active users in a given OCDMA network. In principle, the weight w of an OCDMA address code can be increased to reduce the BER for a fixed number of active users in an OCDMA system using an optical orthogonal code (OOC) [1–3]. But use of larger weight results in higher power losses in an OCDMA system. Moreover, using a larger weight causes higher system cost, because more optical delay lines are employed in the OCDMA system and optical $1 \times w$ splitter/$w \times 1$ combiner of a higher weight are required.

To reduce the effect of OMAI, thermal noise and APD detector noise error-correction codes can be used in OCDMA systems. This will permit a choice of lower weight for OCDMA address codes thus reducing the complexity and power loss of the OCDMA encoder/decoder. As will be explained later, an error correction code is used before OCDMA encoding is done at each transmitter and after OCDMA decoding is performed at each receiver. A well known result from information theory is that randomly chosen code of sufficiently large block length n (or the constraint length in case of convolutional codes) is capable of approaching channel capacity. *Berrou* introduced a new class of error correcting codes called "Turbo codes" which offer a substantial coding gain [4]. They are *parallel-concatenated convolutional codes* (PCCC) whose encoder is formed by two (or more) constituent systematic convolutional encoders joined through a pseudo-random interleaver. Due to the use of pseudo-random interleaver, turbo codes appear random to the channel, yet posses enough structure so that decoding can be physically realized. The decoding is not maximum likelihood (ML) decoding, but tries to approach ML decoding in an iterative way. For the turbo decoding, MAP (*maximum a posteriori*) is known to be an optimal choice. There are many suboptimal algorithms such as SOVA (soft output Viterbi algorithm) and Max-log-MAP, which are less complex than the MAP algorithm. In Trellis coded modulation, coding and modulation are combined together [5]. Redundancy is introduced by using more signal points in the constellation than is required for the modulation format of interest with the same data rate [6]. Convolutional coding is used to introduce a certain dependency between successive signal points. Soft-decision decoding is performed at the receiver, in which the permissible sequence of signals is modeled as a trellis structure.

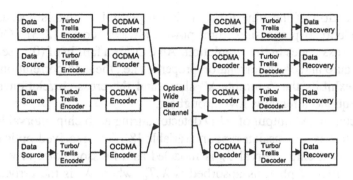

Figure 1. OCDMA Network using a Turbo/Trellis code.

2. SYSTEM MODEL

Multiple accessing is achieved by having multiple sources, each with its own code sequence (called address code), and superimposing their transmissions over a common channel. At the receiver end of the OCDMA system, the optical pulse sequence is compared to a stored replica of itself (correlation process). The correlated value is then compared with a threshold level for data recovery. In an incoherent OCDMA network using optical processing, the data messages at the active transmitters using an on-off key are first encoded with their desired OCDMA address code words and are then distributed to each receiver.

Fig. 1 shows the block diagram of a Turbo/Trellis-coded OCDMA system, where information from each user is first encoded into a turbo/ trellis code by a turbo/ trellis encoder, which is further encoded with the desired OCDMA address code at respective transmitters and then distributed to each receiver. At the receiver side, an OCDMA decoder first decodes the data and then data is fed to a turbo/ trellis decoder to retrieve the original information sent by the user. Each transmitter is assigned a unique codeword from an OCDMA address code. It is assumed that all the optical sources at transmitters are incoherent so that optical power signals of multiple users occurring at the same time would incoherently add in intensity at an OCDMA decoder.

Further identical data rates and signal formats are assumed for all the users and the same effective average power is assumed at the input of each receiver so that one user should not overwhelm the others [4]. In OCDMA systems, each data bit '1' from a source is transformed into the desired destination codeword by using an OCDMA encoder. No light is actually transmitted when each data bit '0' is issued by the data source. The BER performance of OCDMA systems is degraded mainly due to OMAI, which comes from all other active users. At the receiver, the effects of thermal noise and APD noise have been considered. The binary bit '0' might be mistaken for a binary '1' if OMAI signals are strong enough to cause a false detection (called 0-error) at the receiver. But a false detection of the binary bit '1' is not possible. This is because, with incoherent optical processing, light powers always add up [2, 3]. OOCs are used as address codes in the simulation. An OOC is a family of $(0,1)$ sequences with good auto and cross correlation properties i.e., the auto-correlation of each sequence exhibits the "thumbtack" shape and the cross-correlation between any two sequences remains low throughout.

The accumulated output of APD detector during each chip interval has been approximated as a Gaussian random variable [8]. The received optical signal intensity over a chip interval T_c is modeled as a Poisson point process. The average number of photons absorbed is $\lambda_s T_c$, where λ_s is the arrival rate of incident photons due to chip '1' transmission in the signature sequence, which can be represented as

$$\lambda_s = \eta P_w / h f$$

where P_w is the received optical power at optical correlator, η is the APD quantum efficiency, h is the Planck's constant, and f is the optical carrier frequency.

The mean and variance of the conditional probability density function of the accumulated output of APD over the last chip interval can be expressed as:

$$\mu = GT_c[\epsilon\lambda_s + I_b/e] + T_c I_s/e$$
$$\sigma^2 = G^2 F_e T_c[\epsilon\lambda_s + I_b/e] + T_c I_s/e + \sigma_{Th}^2$$

where G is the average APD gain, I_s is the APD surface leakage current, e is the electron charge, ϵ is the total number of marks at the output of the second optical hard limiter and is equal to w if $m = w$ or 0 otherwise, m being the number of non-zero elements, $F_e = K_{eff}G + (2 - 1/G)(1 - K_{eff})$ is the excess noise factor, K_{eff} being the APD effective ionization ratio, $\sigma_{Th}^2 = 2K_B T_r T_c/(e^2 R_L)$ is the variance of the thermal noise, T_r being the receiver noise temperature, K_B the Boltzmann's constant, and R_L the receiver load resistance.

3.   SIMULATION DETAILS

In this paper, we have analyzed, simulated and compared the performance of an uncoded OCDMA system (without error control coding), Turbo-coded OCDMA system and Trellis-coded OCDMA system. The conclusions derived are entirely based on the results of the simulations carried out for 10 active users. It is assumed that errors introduced are due to OMAI, thermal noise and APD noise. Simulation of OCDMA systems required the construction of OOCs. In addition, Turbo coded OCDMA systems required the construction of turbo encoding and turbo decoding functions whereas the Trellis coded modulation systems required trellis transmitter and receiver structures.

The simulation was carried out under the following assumptions:

1. The communication between the transmitters and receivers is pair wise.

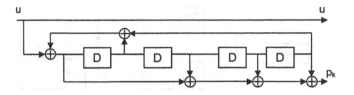

Figure 2. Turbo encoder.

2. Transmitter-receiver pair # 5 is the pair actually sending data and receiving data.

3. All other users send data bits that are randomly generated. Thus, the OMAI effect of all other users on transmitter-receiver pair # 5 is considered.

4. The effects of thermal noise and APD noise are considered.

5. The various transmitter-receiver pairs send data synchronously with respect to each other.

The data being sent by users other than user # 5 is assumed to be truly random. Therefore, with all specifications being the same, if the simulation is repeated, the bit error rate is bound to be different. Hence, to arrive at a generalized value of bit error rate for an OCDMA system with a particular set of specification, numbers of simulations were carried out and average results taken into consideration.

The generator matrix of recursive systematic encoder (RSC) employed by the turbo encoder used in the simulation is $G_R(D) = \begin{bmatrix} 1 & \frac{1+D+D^2}{1+D^2} \end{bmatrix}$ or (1, 7/5, 7/5) in octal, where feed forward generator $g_2(D) = 1 + D + D^2$, feedback generator $g_1(D) = 1 + D^2$. Fig. 2 gives the block diagram of the turbo encoder used in the simulation. The turbo decoder used in the simulation employs Max-logarithmic-MAP algorithm.

The architecture for the transmitter for trellis-coded scheme is given in Fig. 3(a) [6]. The data frame has two slots to represent the state of the input data. If the information bit is same as the preceding digit, then it is encoded into the first slot. Once input data transits from 1 to 0 (or 0 to 1), then it is encoded into the second slot. Hereafter, the precoded data frame is sent to the optical sequence encoder in the upper arm or lower arm for spreading into signature sequence.

If the input data bit is 1, then data frame is transmitted through the upper arm. On the other hand if input data bit is 0, then data frame is transmitted through the lower arm. Thus there are four possible data symbols, and we denote them as α, β, γ and δ respectively. The precoded data symbols are illustrated in Fig. 3(b).

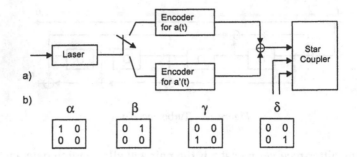

Figure 3.   a) Transmitter architecture; b) Precoded data symbol.

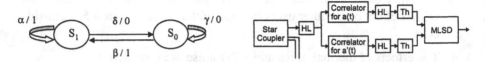

Figure 4.   State transition diagram.   *Figure 5.* Receiver using DHL and MLSD.

If the input data bit is 1, then α is transmitted if previous bit was also 1, otherwise β is transmitted. If the input data bit is 0 then γ is transmitted if previous bit was also 0, otherwise δ is transmitted. Fig. 4 shows the resulting state transition diagram. Each user is assigned, two mutually orthogonal signature sequences generated from time-shifted versions of OOCs.

The correctness of the simulations was tested thoroughly by checking user data and user code words for each bit transmitted. For example: if all other users except user # 5 send only the data bit '0', there will be zero OMAI and number of erroneous bits received should be zero and hence, BER would be zero.

Receiver using double optical hard limiters and maximum likelihood sequence detector (MLSD) is shown in Fig. 5.

4.   SIMULATION RESULTS

BER for various values of threshold for uncoded OCDMA, Turbo coded OCDMA and Trellis coded OCDMA systems are plotted. Fig. 6 compares the BER performance of the uncoded OCDMA system with and without the use of double hard limiter. The system, which includes the double hard limiter, has an improved BER performance than a system without it.

Fig. 7 shows the bit error rate performance versus threshold for an OCDMA system employing optical orthogonal codes without coding, with Turbo coding and with Trellis coding.

From these we can infer that for a fixed weight, all the three systems have minimum BER performance when the threshold at the receiver is equal to the

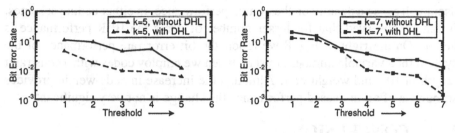

Figure 6. Bit Error Rate versus threshold for OCDMA systems based on (100,5,1,1) (left) ar

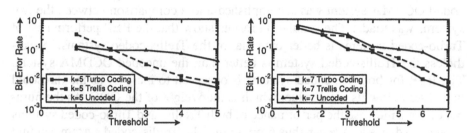

Figure 7. Bit Error Rate versus threshold for OCDMA systems based on (100,5,1,1) (left) and (100,7,1,1) (right) OOC for fixed weight with Turbo Coding and with Trellis Coding.

weight of the codes. A fixed weight Trellis coded or Turbo coded system is better than the uncoded system. Both Trellis and Turbo codes reduce the error floor. The improvement in BER performance can be interpreted as gain due to coding. The decoder of both codes account for the effect of OMAI, thermal and APD noise. The performance of Turbo coded system is found to be better than Trellis coded system. It is also observed that the system with higher code weight can yield better performance.

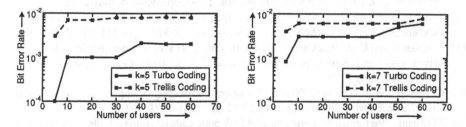

Figure 8. Bit Error Rate versus number of users for OCDMA systems based on (100,5,1,1) (left) and (100,7,1,1) (right) OOC for fixed weight with Turbo Coding and with Trellis Coding.

On the left of Fig. 8 we report the bit error rate performance of the system for a variable number of users, when we employ codes with a constraint length of 100 and weight of 5. Multiple access interference is the major deterioration source for the system performance. It can be seen that an increased number of users will cause a penalty on the system performance, in either coding schemes. It can be observed that for lesser number of users the BER performance is better. On the right of Fig. 8 we report the bit error rate performance of the system for a variable number of users, when we employ codes with a constraint length of 100 and weight of 7. Although the increase in code weight provides somewhat of an increased performance the change is not very significant.

5.   CONCLUSION

The performance analysis of Turbo-coded OCDMA system and Trellis-coded OCDMA system was accomplished and a comparison between the two systems was made. The simulation results show that the BER performance of Turbo-coded systems is better than that of the Trellis-coded systems. Nevertheless the Trellis-coded system is better than the uncoded OCDMA system. The BER for both the coded systems is observed to increase as the threshold decreases below its ideal value, which is the weight of the OOC. As the number of users increase the performance of both Trellis- and Turbo-coded systems appears to degrade. We can thus propose that the Trellis-coded system can find applications in systems where bandwidth is a crucial factor, decoding time has to be kept low and high reliability is not a constraint. Also the hardware complexity of a Trellis-coded OCDMA system is comparatively less as compared to a Turbo-coded OCDMA system. It can be used without much modification of the existing chipsets. It is ideal for LAN applications. Turbo-coded systems should be used where reliability is very important. However, bandwidth used is more than that used for a Trellis system.

REFERENCES

[1] F. R. K Chung, J. A Salehi and V. K. Wei, "Optical Orthogonal codes: Design, analisis and applications," *IEEE Trans. Inform. Theory*, vol. IT-35, pp. 595–604, May 1989.

[2] J. A Salehi, "Code division multiple access techniques in optical fibre networks –Part 1: Fundamental Principles," *IEEE Trans. Commun.*, vol. COM-37, pp. 824–833, Aug. 1989.

[3] J. A Salehi and C. A. Brackett, "Code division multiple access techniques in optical fibre networks — Part 2: System performance analysis," *IEEE Trans. Commun.*, vol. COM-37, pp. 834–842, Aug. 1989.

[4] C. Berrou and A. Glavieux, "Near optimum error correcting coding and decoding: Turbo Codes," *IEEE Trans. Commun.*, vol. COM-44, pp. 1261–1271, Oct. 1996.

[5] J.G Zhang, "Performance improvement of fiber optic code division multiple access systems by using error correction codes," *IEE proceedings*, online no. 19971454, 1997.

[6] M.-Y. Liu and H.-W. Tsao, "Trellis Coded asynchronous optical CDMA systems," *J. Lightwave Technol.*, vol. 19, no. 2, Feb. 2001.

[7] T. Ohisuki, "Performance analysis of direct detection optical asynchronous CDMA systems with double optical hard limiters," *J. Lightwave Technol.*, vol. 15, pp. 452–457, Mar. 1997.

[8] H. M. Kwon, "Optical orthogonal code-division multiple access system — Part 1: APD noise and thermal noise," *IEEE Trans. Commun.*, vol. 42, pp. 2470–2479, July 1994.

III

CHARACTERIZING, MEASURING, AND CALCULATING PERFORMANCE IN OPTICAL FIBER COMMUNICATION SYSTEMS

A METHODOLOGY FOR CALCULATING PERFORMANCE IN AN OPTICAL FIBER COMMUNICATIONS SYSTEM

Invited Paper

C. R. Menyuk, B. S. Marks, and J. Zweck
Department of Computer Science and Electrical Engineering,
University of Maryland Baltimore County
menyuk@umbc.edu

Abstract: In this paper, we outline a methodology for calculating the performance — in particular the bit error rate — in optical fiber communications systems. We propose the development of an interlocking set of mutually-validating models that range from approximate and system-specific models that are computationally rapid to slower and generic models that are highly accurate. Experimental validation plays an important role in the methodology. We describe the progress in our research group toward fulfilling the goals prescribed by the methodology.

Key words: amplifier noise; simulation; bit error rate.

Calculating the performance in optical fiber communications systems in which nonlinearity plays a significant role in transmission is difficult. The difficulty is further enlarged by the complex way in which different modulation formats — such as the return-to-zero, chirped-return-to-zero, and differential-phase-shift-keying — interact with modern-day receivers. The details of the optical filtering, electrical filtering, and internal nonlinearity can significantly impact the performance in even a standard receiver with hard-decision decoding. The use of foward error correction and signal processing further complicates the calculation of the performance.

The basic difficulty stems from nonlinearity. It has been known since the late 19-th century that nonlinearity poses fundamental difficulties because it can lead to chaos and other highly complex behavior [1]. At the same time, the low error rates that are desired, typically on the order of 10^{-15}, make it impossible to use Monte Carlo simulations. Sometimes, in practice, it is possible to split the problem in two by considering just the error rate in the absence of forward error correction and assuming that the error rate is a known function of the error rate prior to correction. However, that is only possible to do accurately when the noise distribution prior to the correction is Gaussian, which is not always a good approximation [2], and when the weight enumerator function of the error-correcting code is known, which is true for Reed-Solomon codes and some other commonly used codes [3], but is not true in general.

Given the importance of this problem, it is not surprising that dozens of papers have been published on it in the past decade. These papers explore a wide variety of approximations that are described as "analytical," "semi-analytical," "semi-empirical," and so on. They are usually aimed at one aspect of the problem while ignoring others. For example, they might focus on nonlinearly-induced self-phase modulation, while ignoring timing jitter and the impact of the receiver, or they may consider the receiver while neglecting any transmission effects. Terms may be dropped from the propagation equations for mathematical convenience with no clear understanding of the physical meaning of the resulting approximation. Validation, if it is done at all, may be done using Monte Carlo computer simulations where it is not possible to obtain enough realizations to really determine the accuracy of the new method, or with experiments where it is unclear whether the inevitable disagreements between theory and experiment exist are due to experimental error, approximations in the theory, or both. Under these circumstances, it is hard to rely upon these theoretical models when new systems are being designed. So, the deluge of papers on this issue continues.

It is our view that a different theoretical methodology is needed, and we describe it here.

First, it is important to carefully distinguish between a theoretical validation using a less approximate but more computationally time-consuming theory and an experimental validation. Neither is sufficient by itself. The purposes of both are different. The former tests whether the approximations that have been made to solve an equation are sufficiently accurate. The latter tests whether the equations that are being solved contain the correct physics. It is our view that any theoretical method must be compared to highly accurate simulations. In the case of bit error ratio (BER) calculations, that implies carrying out Monte Carlo simulations using importance sampling techniques where needed to explore very low BERs [4]. There are many in our field who are still unfamiliar with these techniques, but our view is that they must become familiar with them in order to properly validate the results of reduced models. Ultimately, it is desirable to create an interlocking set of mutually-validating models that span the range from closed form expressions that are highly approximate and apply only to specific systems, but are computationally rapid, to computer simulations that are computationally slower, but are more generic and are in principle highly accurate.

This set of models must be subjected to careful experimental validation. Without this validation, one cannot be certain that the models that have been developed are based on the correct physics. The aim of this validation should be to examine the correctness of the underlying assumptions that are common to the whole set of models, including the least approximate, not to validate the use of approximations that are used to reduce the models.

Second, it is important when making approximations to have a clear understanding of their physical meaning and when they are expected to be valid. This analysis will allow the user to understand at least roughly when the approximations are expected to fail. Purely mathematical approximations are of little use because one cannot rely upon them when changing parameters. This physical understanding should be carefully validated.

We will briefly describe here work to accurately calculate BERs that has been carried out in our research group in recent years. This work exhibits many but not all of the features of the proposed methodology. We finish by describing the work that needs to be done.

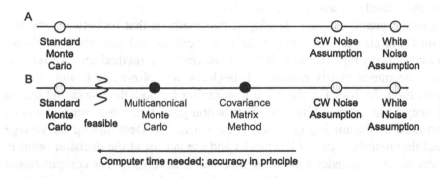

Figure 1. Schematic illustration of different methods for modeling noise-dominated systems. The models range from highly accurate and generic, but computationally intensive on the left to highly approximate and system-specific, but computationally rapid on the right. We show the models that have been historically used (a) without and (b) with the models that we have recently developed.

Historically, the approaches for calculating the BER have been of two types, as shown schematically in Fig. 1a. At one extreme are complete computer simulations based on standard Monte Carlo methods. In principle, with accurate transmitter, fiber, amplifier, and receiver models, one can keep all the nonlinear effects in the problem and calculate the probability distribution functions of the voltage for both the marks and the spaces at the decision point in the receiver. From there, one can in principle determine the optimum threshold voltage in a hard-decision receiver, the minimum BER in any receiver, and other quantities of interest. In practice, one cannot calculate the large number of samples that are needed to accurately determine the distributions. To calculate a BER on the order of 10^{-9} would require on the order of 10^{12} samples, and a BER on the order of 10^{-15} is a more typical target at the present time. Even if one supposes that one is using a forward error correcting code whose properties are known, and that one need "merely" calculate a BERs of 10^{-4} or 10^{-5}, it is still not feasible to calculate accurately using standard Monte Carlo simulations for a broad parameter range. Thus, there is clearly a need for reduced approaches. The most widely used reduced model is to simply ignore the nonlinear interaction between the signal and the noise during transmission and treat the noise as additive, Gaussian, and white at the entry to the receiver. In this case, there is no need for computationally intensive Monte Carlo simulations. A somewhat more advanced approach is to treat the signal as a continuous wave source during transmission [3]. This approach takes into account parametric signal-noise beating. It is known that simplified approaches are sometimes wrong. They disagree with each other and from the means and standard deviations that are calculated using standard Monte Carlo simulations. However, without methods that can quantify the errors, it has not been possible to determine precisely when these methods would be applicable.

In recent years, we have developed two methods that lie between the full Monte Carlo simulations and the reduced methods just described, as shown schematically in Fig. 1b. The first is a deterministic method that we refer to as the covariance matrix method. It neglects the noise-noise beating during transmission. In this case, the noise distribution for both the marks and spaces will obey a multivariate Gaussian distribution prior to the receiver and can be determined by calculating the covariance matrix [5]. Determining the voltage probability distributions for the marks and the spaces at the decision point in the receiver is well-understood for square-law receivers [5]. The computational cost is approximately equal to 200 noise-free simulations in the cases that we studied. The second approach uses multicanonical Monte Carlo simulations [6]. This method is an adaptive importance sampling method and accounts for all the nonlinear effects during transmission and in the receiver. The computational cost is approximately 10^5 noise-free simulations. Both approaches can easily access BERs of 10^{-15}. We have applied both methods to a prototypical undersea system, and the two agree completely. We show the results for the probability distribution functions of the voltages at the decision point in the receiver in Fig. 2.

While neither method is strictly new, both required significant development in order to be applied to optical fiber communications systems with important transmission nonlinearity. The development of these methods for optical fiber communications methods is a significant step in our program of developing

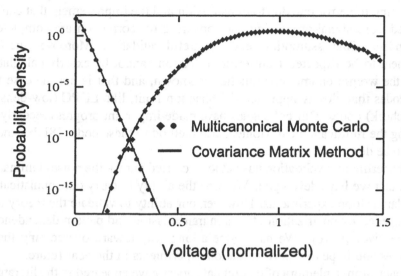

Figure 2.   Comparison of the covariance matrix method to multicanonical Monte Carlo simulations for a prototypical undersea channel system. This figure is taken from [6], and the parameter values are given there.

an interlocking set of methods that will allow one to accurately calculate the performance in these systems. However, there is much that is still lacking. A discussion of the key missing elements follows:

(1) The approaches that we described are appropriate for noise-dominated systems, in which there are no significant pattern dependences among the bits. Nonlinearly-induced pattern dependences in transmission and pattern dependences in the receiver can contribute significantly to the spread of the voltage distribution of the marks and the spaces. Numerous reduced models have been developed in recent years to capture one aspect or another of these effects. Our own group has made progress recently [7] in developing both deterministic and statistical models that can handle these problems and test the simpler and more approximate models that abound in the literature. However, much remains to be done.

(2) In the context of noise-dominated systems, our new models work quite well. However, they are too slow computationally for broad parametric studies. It is an important part of the methodology that we outlined to compare reduced models like the widely-used additive-white-noise-Gaussian model to less approximate models. That will let users know when they can safely use the approximate models and when they are expected to fail.

(3) Receivers are becoming increasingly sophisticated and often contain nonlinear elements, signal processing, and forward error correction. To date,

virtually all the work on BER calculations has assumed that the problem of determining the error rate due to transmission and the improvement that can be expected from signal processing and forward error correction are completely independent. This assumption requires careful validation. Moreover, the improvement to be expected from error correction cannot be exactly calculated unless the weight enumerator function is known, and that is not the case for some codes that closely approach the Shannon limit, like LDPC (low-density parity check) codes. Our colleagues have made important progress recently in applying the multicanonical Monte Carlo method to these codes [8], but more needs to be done.

(4) Experimental validation has yet to be carried out for the noise-dominated models that we have developed. We have the ability to carry out sophisticated recirculating loop experiments. However, our ability to validate the theory has been impeded by polarization effects in transmission and pattern dependences in the receiver [9, 10]. We have gone a long way toward overcoming these difficulties and hope to report validation experiments in the near future.

In conclusion, a plethora of theoretical papers have appeared in the literature in recent years that are aimed at calculating the performance — most particularly the BER — of optical fiber communications systems with important transmission nonlinearity. These models usually cover only one aspect of the problem and are only partially validated if at all. We propose a methodology for calculating performance that calls for the development of an interlocking set of models, from highly reduced models that are approximate and system-specific, but are computationally rapid, to simulation models that in principle contain all the physics, but are computationally slow. These models should cover all aspects of the problem. Experimental validation is an important component of the methodology. Our research group has made important progress towards carrying out the outlined program; however, much remains to be done. The participation of numerous colleagues in creating and experimenting with reduced models has continued and will continue to play a crucial role in the formulation of the set of interlocking models that we have outlined. However, it is our hope that the methodology that we have presented will be broadly adopted and that our colleagues will participate in the development of the whole set of models that is needed.

ACKNOWLEDGMENTS

We are pleased to acknowledge the participation of a number of our current and former students in the work that we cited. These include: R. Holzlöhner, O. Sinkin, Y. Sun, H. Xu, Y. Cai, and A. Mahadevan. The participation of Drs. V. Grigoryan, W. Kath, and J. Morris is also gratefully acknowledged. The work by our research group described here has been supported by the

U. S. Department of Energy, the National Science Foundation, the Laboratory for Physical Sciences, and the Laboratory for Telecommunications Sciences.

REFERENCES

[1] A. J. Lichtenberg and M. A. Lieberman, *Regular and Chaotic Dynamics*. New York, NY: Springer Verlag, 1992.

[2] Y. Cai, J. M. Morris, T. Adalı, and C. R. Menyuk, "On turbo code decoder performance in optical-fiber communication systems with dominating ASE noise," *J. Lightwave Technol.*, vol. 21, pp. 727–734, Mar. 2003.

[3] S. B. Wicker, *Error Control Systems for Digital Communication and Storage*, Upper Saddle River, NJ: Prentice-Hall, 1995.

[4] R. Srinivasan, *Importance Sampling: Applications in Communications and Detection*, New York, NY: Springer-Verlag, 2002.

[5] R. Holzlöhner, V. S. Grigoryan, C. R. Menyuk, and W. L. Kath, "Accurate calculation of eye diagrams and bit error rates in optical transmission systems using linearization," *J. Lightwave Technol.*, vol. 20, pp. 389–400, Mar. 2002. See references cited therein.

[6] R. Holzlöhner and C. R. Menyuk, "Use of multicanonical monte carlo simulations to obtain accurate bit error rates in optical communications systems," *Opt. Lett.*, vol. 28, no. 20, pp. 1894–1896, 2003.

[7] O. V. Sinkin, V. S. Grigoryan, R. Holzlöhner, A. Kalra, J. Zweck, and C. R. Menyuk, "Calculation of error probability in WDM RZ systems in presence of bit-pattern-dependent nonlinearity and of noise," *Proc. OFC 2004*, 2004, paper TuN4.

[8] R. Holzlöhner, A. Mahadevan, C. R. Menyuk, J. M. Morris, and J. Zweck, "Evaluation of the very low BER of FEC codes using dual adaptive importance sampling," *IEEE Comm. Lett.*, to appear.

[9] Y. Sun, I. T. Lima, Jr., A. O. Lima, H. Jiao, J. Zweck, L. Yan, G. M. Carter, and C. R. Menyuk, "System performance variations due to partially polarized noise in a receiver," *IEEE Photon. Technol. Lett.*, vol. 15, pp. 1648–1650, Nov. 2003.

[10] H. Xu, J. Zweck, L. Yan, C. R. Menyuk, and G. M. Carter, "Quantitative experimental study of intrachannel nonlinear timing jitter in a 10-Gb/s terrestrial WDM return-to-zero system," *IEEE Photon. Technol. Lett.*, vol. 16, pp. 314–316, Jan. 2004.

MARKOV CHAIN MONTE CARLO TECHNIQUE FOR OUTAGE PROBABILITY EVALUATION IN PMD-COMPENSATED SYSTEMS

Marco Secondini, Enrico Forestieri, and Giancarlo Prati
Scuola Superiore Sant'Anna di Studi Universitari e Perfezionamento, Pisa, Italy, and Photonic Networks National Laboratory, CNIT, Pisa, Italy.
marco.secondini@cnit.it

Abstract: A novel, fast, accurate and simple method for outage probability evaluation in compensated systems is proposed. The Monte Carlo method is applied to Markov chains to drastically reduce the number of simulations required to estimate rare events. The accuracy is quite independent of the outage probability value to estimate, and the uncertainty can be continuously monitored and reduced at will.

Key words: optical fiber communication; polarization-mode dispersion; Markov chain Monte Carlo; outage probability.

1. INTRODUCTION

Several techniques have been studied and implemented for the evaluation of the outage probability $P(\mathcal{O})$ of optical systems due to polarization mode dispersion (PMD). In particular, two different approaches can be used: the analytical evaluation of $P(\mathcal{O})$ based on simplified (first- or second-order) PMD models and their related statistics [1, 2], or the estimation of $P(\mathcal{O})$ by averaging a high number of simulations based on a complex all-orders PMD model, such as the random waveplate (RWP) model. Usually, analytical methods are very fast, but not so accurate, whereas simulations require a very long computational time. In particular, standard Monte Carlo (MC) techniques are unfeasible for low values ($P(\mathcal{O}) < 10^{-4}$), when adequate confidence intervals are required. Recently, different MC techniques, based on important sampling (IS) [3], or multicanonical MC (MMC) [4], have been proposed to reduce the number of simulations required. Here we propose a novel approach, based on a Markov chain Monte Carlo (MCMC) technique (see [5] and references therein), for a fast and accurate estimation of $P(\mathcal{O})$ down to extremely low values. The method is based on a one-shot MC integration, as in standard MC, with a straight evaluation of $P(\mathcal{O})$. It doesn't require an iterative procedure for subsequent approximations of the probability density function (pdf), as MMC, reducing the number of simulations required and avoiding stability problems. The accuracy can be increased at will by increasing the number of simulations, and can be continuously monitored while performing the evaluation. The method is compared to the analytical method in [2] and to standard MC in the case of a first-order compensated system, showing an excellent agreement.

2.   MARKOV CHAIN MONTE CARLO METHOD

In this paper, we use the bit-error-rate $P(\mathcal{E})$ as a figure of merit of the system performance, considering $P(\mathcal{E}) = 10^{-12}$ as the threshold value defining the outage condition. Thus, $P(\mathcal{O})$ is the probability that $P(\mathcal{E}) > 10^{-12}$. Letting $E \triangleq \log_{10} P(\mathcal{E})$, $P(\mathcal{O})$ can be evaluated as

$$P(\mathcal{O}) = \int_{-\infty}^{0} f(E)p_E(E)dE \tag{1}$$

where $f(E) = u(E + 12)$ is the outage indicator function, $u(\cdot)$ being the unit step function, and $p_E(E)$ the pdf of E. The fiber is described with high accuracy (all-orders PMD) by means of the RWP model: W waveplates with fixed delay $\Delta\tau$ and W polarization rotators[1], with independent and uniformly distributed random rotation angles $\theta_1, \ldots, \theta_W$. A generic realization is determined by the random variable (r.v.) $\theta = (\theta_1, \ldots, \theta_W)$, and E turns out to be a r.v. itself, $E = E(\theta)$. The model can not be used to analytically evaluate $p_E(E)$, but can be used to draw samples according to this pdf. A standard approach is the MC integration of (1) based on independent samples: N independent trials are performed, drawing N independent r.v. $\theta_1, \ldots, \theta_N$, and the RWP model is used to evaluate the corresponding r.v. $E_1, \ldots, E_N$, distributed according to $p_E(E)$. Then, $P(\mathcal{O})$ is estimated by the average

$$\widehat{P}(\mathcal{O}) = \frac{1}{N} \sum_{i=1}^{N} f(E_i). \tag{2}$$

The strong and weak law of large numbers ensure that the estimate in (2) can be made arbitrarily accurate by increasing N, thus requiring a high number of

---

[1]In some cases, and also in this paper, polarization scramblers, rather than simple rotators, are placed between the waveplates. Two parameters are used for each section, and their pdf is such to obtain a uniform distribution of the PMD versor on the Poincaré sphere. For the sake of simplicity, here we describe the simpler model with rotators.

simulations. Indeed, to estimate $P(\mathcal{O})$ from the rate of outage events, several events must be observed. Thus, N is inversely proportional to $P(\mathcal{O})$ itself. In practice, standard MC is feasible only for $P(\mathcal{O}) \geq 10^{-4}$. To overcome this limitation, some techniques have been proposed, such as IS [3] or MMC [4]. In particular, MMC is based on an iterative procedure to build subsequent approximations of $p_E(E)$. Each iteration, a Markov chain is generated, and the samples are used to build a histogram-based approximation of $p_E(E)$. This approximation is used in the next iteration to build a new Markov chain and refine the approximation. Finally, $P(\mathcal{O})$ is evaluated by integrating $p_E(E)$. Our very simple idea is to use Markov chains only to force the model to generate events according to a modified pdf, $p_E^*(E)$, with a higher probability of outage events. In this case, a one-shot MC integration can be performed, and $P(\mathcal{O})$ can be straight estimated by the samples mean, as in (2), without going through the estimation of $p_E(E)$. The increased number of outage events in $p_E^*(E)$ reduces the number of simulations required. Obviously, (2) must be consequently modified to compensate for this artificial increase. The modified pdf can be obtained by generating a Markov chain, $E_1, \ldots, E_N$, with the desired limiting distribution $p_E^*(E)$ by means of a rejection algorithm [5]. In particular, we choose

$$p_E^*(E) = C p_E(E) e^{E/T} \tag{3}$$

where C is a normalization constant, and T is a parameter determining the probability enhancement of outage events, playing the same role as the "temperature" in simulated annealing, to which MCMC is closely related. In this case, the Metropolis rejection algorithm [6] can be used to generate the Markov chain. A generic sample, $E_i = E(\theta_i)$, is obtained from the previous one, $E_{i-1} = E(\theta_{i-1})$, by means of a small random perturbation $\Delta\theta$. The provisional sample, $\widetilde{E}_i = E(\widetilde{\theta}_i)$, with $\widetilde{\theta}_i = \theta_{i-1} + \Delta\theta$, is evaluated, and is accepted ($\theta_i = \widetilde{\theta}_i$ and $E_i = \widetilde{E}_i$) with a probability

$$\min\left(1, e^{\Delta E/T}\right) \tag{4}$$

where $\Delta E = \widetilde{E}_i - E_{i-1}$, and rejected ($\theta_i = \theta_{i-1}$ and $E_i = E_{i-1}$) otherwise. The perturbation $\Delta\theta$ is a vector of W independent Gaussian r.v., with zero mean and variance σ_θ^2/W, generating a random walk in the θ-space, with an average step σ_θ. Different pdfs for $\Delta\theta$, such as a uniform pdf, could be used as well. Under standard conditions for Markov chains, the dependence of the generic realization E_i on any fixed number of early realizations $E_1, \ldots, E_M$ disappears as $i - M \rightarrow \infty$, and the distribution of E_i approaches a stationary form $p_E^*(E)$. To demonstrate that $p_E^*(E)$ converges to the desired pdf in (3), we first analyze a simpler case, when (4) is not applied. In that case, the random perturbation $\Delta\theta$ generates a pure random walk, the generic configuration θ_i has a uniform limiting distribution and, consequently, E_i has the limiting distribution $p_E(E)$. When the stationary distribution is reached, an equilibrium equation holds

$$p_E(E + \Delta E \mid E)p_E(E) = p_E(E \mid E + \Delta E)p_E(E + \Delta E) \qquad (5)$$

In (5), $p_E(E + \Delta E \mid E)$ is the probability that the random perturbation causes the transition $E \rightarrow E + \Delta E$, whereas $p_E(E \mid E + \Delta E)$ is the probability that it causes the opposite transition. Hence, for $\Delta E > 0$, $p_E(E + \Delta E \mid E)p_E(E)$ is the probability of an upward transition, whereas $p_E(E \mid E + \Delta E)p_E(E + \Delta E)$ is the probability of a downward transition. Eq. (5) means that, once the stationary distribution is reached, downward and upward transitions must be equiprobable. When applying (4) to the Markov chain generation, an analogous equilibrium equation holds

$$p_E^*(E + \Delta E \mid E)p_E^*(E) = p_E^*(E \mid E + \Delta E)p_E^*(E + \Delta E) \qquad (6)$$

The conditional probabilities in (6) can be evaluated considering that the probability to get state $E + \Delta E$, starting from state E, or vice versa, is equal to the probability that the random perturbation causes the transition, multiplied by the probability that the transition is accepted, obtaining

$$p_E^*(E + \Delta E) = p_E^*(E)\frac{p_E(E + \Delta E)}{p_E(E)}e^{\Delta E/T} \qquad (7)$$

By using (7) and applying L'Hôpital's rule, we can write a differential equation for $p_E^*(E)$

$$\frac{dp_E^*(E)}{dE} = \lim_{\Delta E \to 0}\frac{p_E^*(E + \Delta E) - p_E^*(E)}{\Delta E} = \frac{p_E^*(E)}{p_E(E)}\left[p_E'(E) + \frac{1}{T}p_E(E)\right] \qquad (8)$$

where the prime denotes derivative with respect to E. Integrating (8), we obtain

$$p_E^*(E) = p_E^*(E_0)e^{\frac{E - E_0}{T} + \int_{E_0}^{E}\frac{p_E'(\epsilon)}{p_E(\epsilon)}d\epsilon} = Cp_E(E)e^{E/T} \qquad (9)$$

where all constant values have been grouped into the constant C. This proves that the limiting distribution of the Markov chain is our desired pdf (3). Using the modified pdf, we can rewrite (1) as

$$P(\mathcal{O}) = \frac{1}{C}\int_{-\infty}^{0}f(E)e^{-E/T}p_E^*(E)dE \qquad (10)$$

and, since the ergodic theorem allows us to substitute the expected value with the time average over the sequence, as in standard MC we can estimate (10) from the samples mean

$$\widehat{P}(\mathcal{O}) = \frac{1}{NC}\sum_{i=1}^{N}f(E_i)e^{-E_i/T} \qquad (11)$$

Similarly, the normalization constant C can be estimated as

$$\widehat{C} = \frac{1}{N} \sum_{i=1}^{N} e^{-E_i/T}. \tag{12}$$

3.   ACCURACY

When dealing with MC methods, the very important question of how much simulations are needed to reach a satisfactory accuracy arises. Notice that $\widehat{P}(\mathcal{O})$ itself in (2) or (11), is a r.v., and as in standard MC the samples E_i, $i = 1, \ldots, N$, in (2) are independent, the variance of the estimate is

$$\sigma^2_{P(\mathcal{O})} \cong \frac{\widehat{P}(\mathcal{O})}{N} \tag{13}$$

Defining the relative uncertainty δ as the ratio between the standard deviation $\sigma_{P(\mathcal{O})}$ and $\widehat{P}(\mathcal{O})$ itself

$$\delta = \frac{\sigma_{P(\mathcal{O})}}{\widehat{P}(\mathcal{O})} \cong \frac{1}{\sqrt{N\widehat{P}(\mathcal{O})}} \tag{14}$$

we see that δ can be arbitrarily decreased by increasing the number of simulations N. From (14), for a fixed uncertainty δ, N is inversely proportional to $\widehat{P}(\mathcal{O})$, making the method unfeasible for too low values. For instance, only 10^4 simulations are needed to estimate $P(\mathcal{O}) = 10^{-2}$ with a 10% uncertainty, but quite 10^8 simulations to estimate $P(\mathcal{O}) = 10^{-6}$ with the same uncertainty. Instead, in the proposed method the uncertainty is strongly reduced by using the modified distribution $p_E^*(E)$, and N is quite independent of $P(\mathcal{O})$. However, two counter-factors pose a limit to the method: the normalization constant C has to be estimated too, and the samples $E_1, \ldots, E_N$ in (11) are not independent. In particular, the higher the correlation between the samples, the lower the statistical importance of each sample and the higher the variance of the estimate. Moreover, the correlation makes more difficult to estimate the uncertainty. From (11) and (12), the estimate can be written as

$$\widehat{P}(\mathcal{O}) = \frac{\widehat{O}}{\widehat{C}} = \frac{\frac{1}{N}\sum_{i=1}^{N} f(E_i)e^{-E_i/T}}{\frac{1}{N}\sum_{i=1}^{N} e^{-E_i/T}} \tag{15}$$

that is, the ratio between the estimated outage term $\widehat{O}$ and normalization term $\widehat{C}$. The variance of the ratio can be estimated as

$$\sigma^2_{P(\mathcal{O})} \cong \frac{\sigma^2_O - 2\widehat{P}(\mathcal{O})\sigma_{O,C} + \widehat{P}(\mathcal{O})^2\sigma^2_C}{\widehat{C}^2} \tag{16}$$

where σ_O^2, σ_C^2 and $\sigma_{O,C}$ are the variance of $\widehat{O}$, the variance of $\widehat{C}$ and their covariance, respectively, and can be estimated as suggested in [7]. The samples are subdivided into L groups of K consecutive samples each. The sample means $\overline{O}_i$ and $\overline{C}_i$, for $i = 1, \ldots, L$, are evaluated for each group, and σ_O^2, σ_C^2 and $\sigma_{O,C}$ are estimated as

$$\sigma_O^2 \cong \frac{1}{L(L-1)} \sum_{i=1}^{L} \left(\overline{O}_i - \widehat{O}\right)^2$$

$$\sigma_C^2 \cong \frac{1}{L(L-1)} \sum_{i=1}^{L} \left(\overline{C}_i - \widehat{C}\right)^2$$

$$\sigma_{O,C} \cong \frac{1}{L(L-1)} \sum_{i=1}^{L} \left(\overline{O}_i - \widehat{O}\right) \left(\overline{C}_i - \widehat{C}\right) \tag{17}$$

The uncertainty can not be predicted a priori, as in standard MC, but can be continuously monitored while performing the simulations. Thus, the desired accuracy can be fixed, and the simulations stopped when it is reached.

The method is very simple and can be straightforwardly implemented. Only two parameters have to be adjusted: T and the average step σ_θ. Their optimization is not critical, but some guidelines can be given to reduce the number of simulations required. In particular, the parameter T is used to increase the probability of outage events: the lower T, the higher the probability of an outage event. A too high T value makes the outage events improbable and, in the limit $T \to \infty$, the distribution $p^*(E) \to p(E)$, and the method becomes a standard MC. Otherwise, a too low T value makes the non-outage events improbable, causing the opposite problem. Thus, T should be chosen to balance the number of outage and non-outage events. For instance, if we want to estimate $P(\mathcal{O})$ in the range $10^{-5} \div 10^{-7}$, a good choice for T is such that $e^{\Delta E_0 / T} = 10^6$, where ΔE_0 is the difference between the most probable value of E (the back-to-back value can often be used), and the outage limit ($E = -12$). On the other hand, the average step σ_θ plays a fundamental role to control the degree of correlation between the samples of the Markov chain. When a step is accepted, the correlation between the previous and next sample depends on the variation of the configuration $\Delta\theta$. Thus, their correlation decreases for an increasing σ_θ. On the contrary, when a step is refused, the next sample coincides (and thus is fully correlated) with the previous one. Hence, their correlation increases for an increasing σ_θ, since the higher the average step, the higher the probability that a step will be refused. Since a high correlation between the samples reduces their statistical importance, the correlation should be minimized in order to minimize the uncertainty. Thus, σ_θ should be chosen to balance the correlation induced by accepted and refused steps. As a

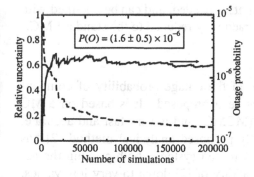

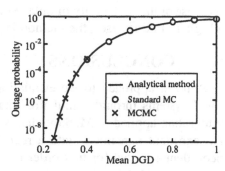

Figure 1. Outage probability and relative uncertainty vs. number of simulations.

Figure 2. Outage probability vs. mean DGD in a first-order compensated system.

matter of fact, we found that the optimization of σ_θ is not critical, and a value in the range $1 \div 10$ can be used in most practical cases.

4. RESULTS

The proposed method has been used to estimate the outage probability of a system with a 3-dB OSNR margin and first-order PMD compensation. For a given mean DGD, the following very easy procedure has been followed:

- Definition of the RWP model to be used (number of waveplates, ...).

- Choice of parameter T and average step σ_θ.

- Generation of the Markov chain according to (4), with continuous update of $\widehat{P}(\mathcal{O})$ and δ according to (15)–(17).

- Stop when the desired uncertainty is reached.

As an example, in Fig. 1 $\widehat{P}(\mathcal{O})$ and δ are reported as a function of the number of samples N, for a mean DGD equal to 30% the bit-time. The RWP model was used with 100 waveplates with fixed DGD and polarization scramblers. Each simulation, E was analytically evaluated as in [8]. In this case, following the guidelines in the previous section, we chose $T = 0.7$ and $\sigma_\theta = 3$. A fast convergence of $\widehat{P}(\mathcal{O})$ can be observed. About 200000 samples are needed to achieve a 10% relative uncertainty, but only 50000 to achieve a 20% relative uncertainty. Considering a confidence range of $\pm 3\sigma_{P(\mathcal{O})}$, we can pose $P(\mathcal{O}) = (1.6 \pm 0.5) \times 10^{-6}$. Finally, in Fig. 2 $\widehat{P}(\mathcal{O})$ is plotted as a function of the mean DGD, and three different evaluation methods are compared: the analytical method in [2], a standard MC method (high values), and the proposed method (low values). The excellent agreement is twofold valuable: it proves the accuracy of the proposed method, and extends the validation of the analytical one down to very low OP values. For high values of $P(\mathcal{O})$, only a little increase of the probability of outage events is needed, and can be obtained with a high T. In this cases, the method is practically reduced to the standard MC.

5. CONCLUSIONS

A novel approach to the evaluation of the outage probability of compensated systems through simulations has been proposed. It is based on a MC integration applied to Markov chains (MCMC) and is very fast and accurate. An excellent agreement with a recently proposed analytical method [2] has been demonstrated for first-order compensated systems, both proving the accuracy of MCMC and extending the validity of [2] down to very low values. With respect to the analytical method, the proposed MCMC is based on an all-orders PMD model and can be used regardless of the complexity of the PMD compensator. Moreover, with respect to standard MC, it allows a fast and accurate evaluation down to extremely low values. Finally, with respect to iterative MC procedures, such as MMC, it is based on a one-shot MC integration and on the straight evaluation of $P(\mathcal{O})$, without the need of an histogram-based pdf estimation, and without any problem related to the stability of the iterative procedure. Thus, the accuracy of the method can be increased at will. A procedure for the continuous monitoring of the uncertainty has been described as well, allowing to stop the integration when the desired accuracy is reached.

REFERENCES

[1] H. Bülow, "System outage probability due to first- and second-order PMD," *IEEE Photon. Technol. Lett.*, vol. 10, pp. 696–698, May 1998.

[2] E. Forestieri and G. Prati, "Exact analytical evaluation of second-order PMD impact on the outage probability for a compensated system," *J. Lightwave Technol.*, vol. 22, pp. 988–996, Apr. 2004.

[3] S. L. Fogal, G. Biondini, and W. L. Kath, "Multiple importance sampling for first- and second-order polarization-mode dispersion," *IEEE Photon. Technol. Lett.*, vol. 14, pp. 1273–1275, Sep. 2002.

[4] A. O. Lima, J. I. T. Lima, J. Zweck, and C. R. Menyuk, "Efficient computation of PMD-induced penalties using multicanonical monte carlo simulations," in *Proc. ECOC'03*, 2003. Paper We3.6.4.

[5] J. C. Spall, "Estimation via Markov chain Monte Carlo," *IEEE Contr. Syst. Mag.*, pp. 34–45, Apr. 2003.

[6] N. Metropolis, A. W. Rosenbluth, M. N. Rosenbluth, A. H. Teller, and E. Teller, "Equation of state calculations by fast computing machines," *J. Chem. Phys.*, vol. 21, no. 6, pp. 1087–1092, 1953.

[7] W. K. Hastings, "Monte Carlo sampling methods using Markov chains and their applications," *Biometrika*, vol. 57, pp. 97–109, 1970.

[8] E. Forestieri, "Evaluating the error probability in lightwave systems with chromatic dispersion, arbitrary pulse shape and pre- and postdetection filtering," *J. Lightwave Technol.*, vol. 18, pp. 1493–1503, Nov. 2000.

A PARAMETRIC GAIN APPROACH TO PERFORMANCE EVALUATION OF DPSK/DQPSK SYSTEMS WITH NONLINEAR PHASE NOISE

P. Serena, A. Orlandini, and A. Bononi
Università degli Studi di Parma, Dipartimento di Ingegneria dell'Informazione
Parco area delle scienze 181/A, 43100 Parma (Italy)
serena@tlc.unipr.it

Abstract: We present a novel method based on parametric gain (PG) to study the impact of nonlinear phase noise in dispersion-managed differentially phase-modulated optical transmission systems. By linearizing the interaction of signal and noise, the received ASE is approximated as a stationary Gaussian process, whose statistics are found by using a low-pass filtered version of the modulating signal. The BER is then evaluated by adapting standard methods for quadratic detectors, for single-channel binary and quaternary PSK systems, both for NRZ and RZ supporting pulses. We show that in the RZ case parametric gain causes a larger penalty than in the NRZ case, and that DQPSK is less robust to PG than DPSK.

Key words: phase noise; differential phase-shift keying (DPSK); differential quadrature phase-shift keying (DQPSK); parametric gain; Gaussian noise.

1. INTRODUCTION

Optical phase shift keying (PSK) modulation formats are an interesting alternative to on-off keying (OOK) for long-haul transmission systems, because of their lower optical signal-to-noise ratio (OSNR) requirements, which leads to reduced power and thus reduced Kerr nonlinearities. PSK formats are usually implemented as differential binary (DPSK) or quaternary (DQPSK) schemes [1, 2]. DPSK encodes the information onto the differential optical phase between adjacent bits which can be either 0 or π, while DQPSK utilizes two orthogonal DPSK signals, which leads to four possible values of the differential phase $0, \frac{\pi}{2}, \pi, \frac{3}{2}\pi$, thus doubling spectral efficiency. However, PSK formats are extremely vulnerable to nonlinear phase noise [3] arising by the nonlinear interaction of signal and amplified spontaneous emission (ASE) noise during propagation. Such an interaction manifests itself also as a parametric gain (PG) of the received ASE noise. It has been experimentally shown that, in 10 Gb/s DPSK systems, the statistics of the decision variable in presence of nonlinear phase noise develop a much larger skewness than without nonlinearity, and resemble an exponential distribution [1]. Attempts have been made to theoretically study the statistics of the nonlinear phase noise (and an exact expression has been found only at zero group-velocity dispersion (GVD) [4]), in order to assess the performance of DPSK receivers based on ideal phase discriminators.

In this work we instead address the performance of realistic PSK receivers based on Mach-Zehnder delay demodulators. Although the statistics of the signal-inflated ASE noise depart from Gaussian at either very large signal levels or at zero GVD, we first show that in practical systems, working at sufficiently large OSNR and in which some local GVD is present, the received ASE noise can still be reasonably described by a Gaussian process.

While with non-return to zero (NRZ) supporting pulses the intensity is constant and the received ASE is a stationary process, in presence of return-to-zero (RZ) pulses the ASE parametric gain becomes time dependent, and thus the ASE is a non-stationary process. The exact non-stationary statistics of the ASE can in principle be obtained through a computationally heavy method based on the linearized solution of the nonlinear Schrödinger equation (NLSE) [5]. In this work, we find instead an *equivalent* stationary ASE process yielding the correct statistics of the decision variable. The bit-error rate (BER) is finally evaluated by a suitable modification of well-established methods for quadratic detectors in Gaussian noise [6].

We apply our model to provide a comparative study of the performance of single-channel dispersion managed DPSK and DQPSK systems, both for NRZ and RZ supporting pulses. Results are provided in terms of OSNR penalties for different system bitrates, in-line residual dispersion and cumulated nonlinear phase.

2.   SYSTEM SET-UP

In Fig. 1 we schematically describe the single channel dispersion-managed DPSK multi-span system studied in the following sections. Before and after the transmission link, pre- and post- compensating fibers are optimized in presence of PG for each launched power value, while all spans have identical residual dispersion, varied in a suitable range at different bitrates. The receiver has an optical Gaussian filter of bandwidth $B_0 = 1.8R$, being R the system bitrate, followed by a Mach-Zehnder interferometer with one bit delay equal to $T = 1/R$. Half the sum of the input fields at times t and $t - T$ is detected by one photodetector, while half the difference is detected by the other. The difference between the received currents is then filtered by a Butterworth 2nd order filter, of bandwidth $B_e = 0.65R$, and finally sampled. For the DQPSK

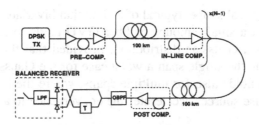

Figure 1. Set-up of the single channel dispersion managed DPSK system.

format, we consider two independent DPSK demodulators for the detection of the in-phase and quadrature signal components, with an additional delay of $\pi/4$ and $-\pi/4$ in each of the Mach-Zehnders [2]. We approximate the DQPSK in-phase and quadrature components as independent random variables (RVs), so that the BER is the average between the BER on each of such components.

3. ASE STATISTICS

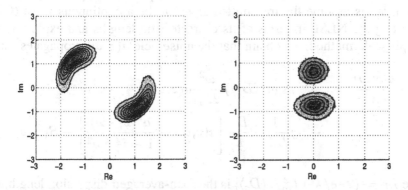

Figure 2. Contour plots of ASE PDF before the receiver for OSNR=25 dB, $\Phi_{NL} = 0.3\pi$ rad and $D_{tx} = 0$ ps/nm/km (left) or $D_{tx} = 4$ ps/nm/km (right).

It is known that signal and ASE have maximum nonlinear interaction strength at zero GVD, so that in such a case the received ASE has statistics far from Gaussian. We estimated the joint probability density function (PDF) of the in-phase (real) and quadrature (imaginary) received ASE components through a Monte Carlo simulation of a single fully-compensated fiber span with an NRZ-DPSK signal (see Fig. 1) operating at a received OSNR of 25 dB (over 0.1 nm), with a transmission fiber chromatic dispersion D_{tx} equal to either 0 or 4 ps/nm/km and for an average cumulated nonlinear phase $\Phi_{NL} = 0.3\pi$ rad. At $D_{tx} = 0$ we obtained the joint PDF contour levels shown in Fig. 2 (left), which confirm that Kerr nonlinearity alone causes a non Gaussian ASE. When including fiber GVD, we obtained the contour levels shown in Fig. 2 (right), which have an elliptical shape, typical of a Gaussian bivariate distribution. We conclude that even a small amount of fiber GVD tends to reshape the ASE joint PDF towards a Gaussian distribution, provided that the OSNR is sufficiently large. Being the single span a worst case for the Gaussian assumption, such a conclusion holds also for multiple spans, where the increased number of independent noise sources accelerates the convergence to a Gaussian PDF.

4.   MODEL FOR NOISE PROPAGATION

In this section we calculate the ASE noise statistics starting from a linearized solution of the NLSE [5] applied to long *periodic* dispersion-managed systems. For such systems the local dispersion is a z-periodic function, being z the distance, which can be written as $D_c(z) = \langle D_c \rangle + L^{-1}\Delta D_c(z/L)$, where $\langle D_c \rangle$ is the span-averaged dispersion, i.e. the in-line residual dispersion per unit length, and $L^{-1}\Delta D_c(z/L)$ is the local deviation from the average, with L equal to the single span length. In this context, the dynamic behavior of the propagating field can be described by a "slow" length scale z and a "fast" scale z/L, which allows to study the NLSE by a multiple scale approach [7].

First, let us assume the transmitted signal to be a continuous wave (CW). By writing the NLSE in terms of its characteristic lengths and exploiting the multiple scale method, we obtain that the noise field $a(z,\omega)$ propagates as:

$$\frac{\partial a(z,\omega)}{\partial z} = + j \, \text{sgn}\left(\langle D_c \rangle\right) \frac{\omega^2}{2L_D} a(z,\omega)$$

$$- j\frac{1}{L_{NL}} \cdot \frac{L_A}{L}\left[a(z,\omega) + \frac{a^*(z,-\omega)}{1 - j\frac{L_A}{L_\Delta}\omega^2}\right] + S(z,\omega) \quad (1)$$

where $L_D = (2\pi c/\lambda^2)\,T_m^2/\,|\langle D_c \rangle|$ is the span-averaged dispersion length, being λ the signal wavelength and T_m the supporting pulse duration. The differential dispersion length $L_\Delta = (2\pi c/\lambda^2)\,T_m^2/(D_{tx} - \langle D_c \rangle)$ accounts for the deviation of the dispersion length $L_d = (2\pi c/\lambda^2)\,T_m^2/\,|D_{tx}|$ of the transmission fiber from the span average value, i.e. $1/L_d = 1/L_D + 1/L_\Delta$. L_A is the attenuation (or effective) fiber length, L_{NL} is the transmission fiber nonlinear length, proportional to the inverse of the signal power P. All the dispersive lengths are referred to $T_m = d \cdot T$, being d the signal duty-cycle, and ω is the frequency normalized to $1/T_m$. $S(z,\omega)$ is a Langevin Gaussian forcing term with autocorrelation at times t_1, t_2 equal to $\langle S(z_1,t_1)\,S^*(z_2,t_2)\rangle = \sigma^2\delta(z_1 - z_2)\,\delta(t_1 - t_2)$, where σ^2 is the white ASE power spectral density (PSD) per unit length and $\delta(\cdot)$ is the Dirac delta function. Eq. (1) is a linear differential equation with constant coefficients which can be cast in closed form in terms of a matrix exponential. For a given distance z, this solution scales with the dimensionless parameters z/L_D, L_A/L_Δ and $\Phi_{NL} = (z/L) \cdot (L_A/L_{NL})$.

Being $S(z, \omega)$ a Gaussian stationary noise, $a(z, \omega)$ is again a stationary Gaussian stochastic process, whose PSD can be obtained in a closed form.

In the DPSK modulated case, the nonlinear length L_{NL} is now a function of time, and eq. (1) does not apply. In such a case the linearization approach yields a linear time-varying approximation of the NLSE, whose solution still yields a Gaussian, but non-stationary noise. However, eq. (1) reveals that, at a specific ω, the noise field depends only on the neighboring frequency samples within a proper bandwidth, which is essentially set by L_A/L_Δ and slightly varies with the CW power P. Such a bandwidth corresponds to a finite memory window in the time domain. Hence, we expect that signal modulation adds only a small perturbation to eq. (1), which can still be used for evaluating the noise statistics at any time t by substituting the signal power $P(z, t)$ with a slowly varying, low-pass filtered version of it $P_{eff}(z, t)$, obtained by means of a windowed Fourier transform. We found that a proper filtering window is $H(\omega) = 1/\left[1 + \left(L_A \omega^2/(4L_\Delta)\right)^2\right]$. Whenever PG is the main impairment to system performance, we assume that the signal has no inter-symbol interference (ISI), i.e. $P(z, t) \simeq P(0, t)$. Hence, at each sampling time t_k of a RZ-DPSK signal with sinusoidally-varying intensity profile, the noise PSD can be evaluated by solving (1), where L_{NL} is calculated with the effective power value:

$$P_{eff}(t_k) = \frac{1}{2}\left(1 + H(\pi)\right) P_{peak} \qquad (2)$$

where P_{peak} is the peak power, while for NRZ-DPSK P_{eff} coincides with P_{peak} (which is also the average power). By this way, the noise can be assumed as stationary for BER computation. Note that, since the noise PSD can be obtained in a closed form, our solution can then be applied to a fast BER algorithm, which avoids the computational burden of the exact method proposed in [5].

5. RESULTS AND DISCUSSION

We first tested our model by trying to replicate the experimental results in [1]. Fig. 3 shows the Q-factor penalty measured in [1] with circles and the prediction of our model (solid line) with NRZ- (left) and RZ-DPSK (right) ($d = 0.33$), for a launched power of 7 dBm. For the system parameters see [1]. For the RZ case, we plot the Q penalty obtained by using either the average P_{avg}, or the peak P_{peak}, or the effective power P_{eff} in the ASE PSD evaluation. From the comparison with experimental data, P_{eff} is found to give the best approximation. In Fig. 4 we also numerically tested our BER results obtained with P_{avg} (circles), P_{peak} (diamonds) and P_{eff} (up-triangles) and by comparing them to the exact BER (down-triangles) evaluated with the algorithm proposed in [5]. We studied a 20-span fully compensated RZ-DPSK ($d = 0.5$) system (see Fig. 1), with a GVD transmission fiber $D_{tx} = 2$ ps/nm/km, at $L_A/L_\Delta =$

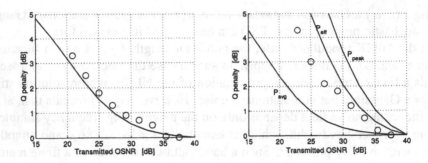

Figure 3.   Q-penalty versus transmitted OSNR for NRZ- (left) and RZ-DPSK (right) for the experimental system tested in ref. [1] with a launched power of 7 dBm.

0.35 (i.e., $R = 40$ Gb/s) and for an average $\Phi_{NL} = 0.3\pi$. The best fit is given by using P_{eff}, while use of P_{avg} or P_{peak} under/overestimates the BER.

Having tested the accuracy of our model for BER calculation, we then applied it in a comparative study among NRZ- and RZ-DPSK and DQPSK system performance (see Section 2), evaluated for varying system parameters.

In Fig. 5 we plot the OSNR penalty (@BER $= 10^{-10}$) versus L_A/L_Δ for $\Phi_{NL} = 0.1\pi$ (triangles), $\Phi_{NL} = 0.3\pi$ (circles) and $\Phi_{NL} = 0.5\pi$ (diamonds) computed with (solid line) and without (dashed line) PG, for NRZ- (left) and RZ-(right) DPSK (top) and DQPSK (bottom) signals. The in-line residual dispersion is optimized with PG. In the NRZ case the performance is set by PG at low L_A/L_Δ, while self-phase modulation (SPM) induced signal distortion is dominant at high L_A/L_Δ. For this reason, the relative penalties between the curves with and without PG decrease for increasing L_A/L_Δ, in both DPSK and DQPSK case. RZ pulse coding ($d=0.5$) is more robust to ISI both with

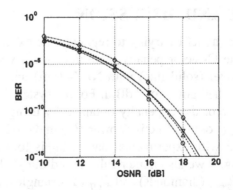

Figure 4.   BER versus OSNR for a 40 Gb/s 20-span full compensated system. Triangles up: exact BER [5]. Triangles down: proposed model by using eq. (2). Circles/Diamond: proposed model with P_{av}/P_{peak}.

and without PG at high L_A/L_Δ, while at low L_A/L_Δ it has worse performance than NRZ coding [1], as expected since P_{peak} is doubled. Note that, at a fixed L_A/L_Δ, the reduced distance among the DQPSK symbols reduces its robustness to SPM distortion. In Fig. 6 we focused on the OSNR penalties plotted versus Φ_{NL} for the same system of Fig. 5, now studied for $D_{tx}=$ 8 ps/nm/km at $R = 40$ Gb/s ($L_A/L_\Delta = 1.4$) and RZ coding. Since DQPSK supports two channels at an halved symbol rate compared to DPSK, it is found to be much less robust to PG, being PG much stronger at lower symbol rates. For the same DPSK system studied in Fig. 6, we investigated the RZ-DPSK dependence on the normalized in-line dispersion per span z/L_D. In Fig. 7 we show the contour plots of the additional OSNR penalty due to PG versus Φ_{NL} and z/L_D. Best performance is found for negative in-line dispersion, where a large tolerance to z/L_D variations is found up to $\Phi_{NL} = 0.25\pi$.

6. CONCLUSIONS

We showed that in dispersion-managed DPSK and DQPSK systems the impact of nonlinear phase noise can be studied by a parametric gain approach which describes the ASE noise with Gaussian statistics, for which we provide a simple and accurate model. Such a model is applied to BER evaluation for

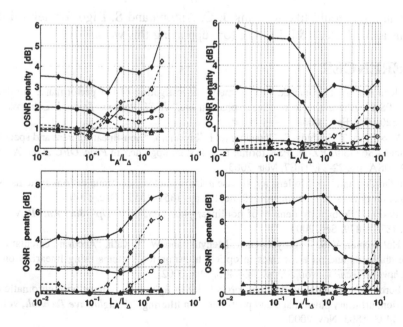

Figure 5. OSNR penalties versus L_A/L_Δ for NRZ- (left) and RZ- (right) DPSK (top) and DQPSK (bottom) modulation formats with (solid line) and without PG (dashed line). Triangles: $\Phi_{NL} = 0.1\pi$; Circles: $\Phi_{NL} = 0.3\pi$; Diamonds: $\Phi_{NL} = 0.5\pi$.

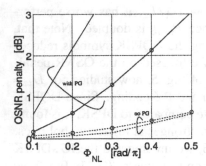

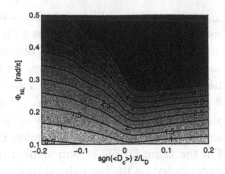

Figure 6. OSNR penalties versus Φ_{NL} for RZ-DPSK (circles) and DQPSK (no symbols) with (solid line) and without PG (dashed line) at $L_A/L_\Delta = 1.4$.

Figure 7. Additional OSNR penalties due to PG for RZ-DPSK at $L_A/L_\Delta = 1.4$ versus Φ_{NL} and $\frac{z}{L_D}$.

quadratic detectors in Gaussian noise. We found that even if RZ coding is more tolerant to SPM distortion than NRZ, it is less robust to PG, whose effect is enhanced by signal modulation. DQPSK doubles the spectral efficiency of DPSK at the price of a reduced tolerance to PG.

ACKNOWLEDGMENTS

The authors would like to thank J.-C. Antona and S. Bigo for helpful discussions and Alcatel R&I, France, for supporting this work.

REFERENCES

[1] H. Kim and A. H. Gnauck, "Experimental investigation of the performance limitation of DPSK systems due to nonlinear phase noise," *IEEE Photon. Technol. Lett.*, vol. 15, pp. 320–322, Feb. 2003.

[2] O. Vassilieva, T. Hoshida, S. Choudhary, H. Kuwahara, "Non-linear tolerant and spectrally efficient 86 Gbit/s RZ-DQPSK format for a system upgrade," in *Proc. OFC 2003*, Atlanta, GA, USA, pp. 22-24, ThE7, Mar. 2003.

[3] J.-P. Gordon and L. F. Mollenauer, "Phase noise in photonic communications systems using linear amplifiers," *Opt. Lett.*, vol. 15, no. 23, pp. 1351–1353 , Dec. 1990.

[4] K. Po-Ho, "Probability density of nonlinear phase noise," *J. Opt. Soc. Am. B*, vol. 20, pp. 1875–1879, Sep. 2003.

[5] R. Holzlohner, V. S. Grigoryan, C. R. Menyuk, and W. L. Kath, "Accurate calculation of eye diagrams and bit error rates in optical transmission systems using linearization," *J. Lightwave Technol.*, vol. 20, pp. 389–400, Mar. 2002.

[6] E. Forestieri, "Evaluating the error probability in lightwave systems with chromatic dispersion, arbitrary pulse shape and post-detection filtering," *J. Lightwave Technol.*, vol. 18, pp. 1493–1503, Nov. 2000.

[7] M. J. Ablowitz, and T. Hirooka, "Resonant intrachannel pulse interactions in dispersion-managed transmission systems," *IEEE J. Quantum Electron.*, vol. 8, pp. 603–615, May-June 2002.

CHARACTERIZATION OF INTRACHANNEL NONLINEAR DISTORTION IN ULTRA-HIGH BIT-RATE TRANSMISSION SYSTEMS

Invited Paper

Robert I. Killey, Vitaly Mikhailov, Shamil Appathurai, and Polina Bayvel
Optical Networks Group, Department of Electronic and Electrical Engineering,
University College London, Torrington Place, London WC1E 7JE, UK
r.killey@ee.ucl.ac.uk

Abstract: Signal distortion due to intrachannel cross-phase modulation and four-wave-mixing limits transmission distances in ultra-high bit-rate optical communications. To gain an understanding of the effects of nonlinear pulse interactions and to quantify the effectiveness of new methods to suppress them, accurate characterization techniques are required to isolate the effects of fibre nonlinearity from the other impairments which occur in transmission. In this paper, we discuss two techniques: firstly, the direct measurements of the signal waveform distortion (pulse timing jitter, amplitude fluctuations, and FWM-induced 'ghost' pulse power) and, secondly, measurements of the BER-dependence on optical signal launch power. We describe the use of these characterization methods to investigate the suppression of nonlinear distortion through the use of optimized dispersion maps, alternate-polarization and alternate-phase return-to-zero signal formats.

Key words: dispersion management; optical fiber nonlinearity; four wave mixing; cross phase modulation.

1.   INTRODUCTION

Limits to the transmission distances achievable in high bit-rate systems are imposed by nonlinear refraction. The intensity modulation of the signals induces optical phase shifts due to the intensity-dependence of the refractive index of the transmission fibres. In WDM systems with narrow channel spacing, the nonlinear refraction leads to interactions between channels through cross-phase modulation (XPM) and four-wave-mixing (FWM). As the channel bit-rate and channel wavelength spacing is increased, the impact of these inter-channel effects is reduced, due to the walk-off between the channels caused by the fibre dispersion, and intrachannel nonlinear effects become increasingly dominant. These effects result from the dispersion-induced pulse broadening and the consequent nonlinear interaction between overlapping pulses within the same channel.

As in the multi-channel case, the pulse overlap leads to cross-phase modulation and four-wave-mixing. These effects were investigated experimentally and numerically in [1–10] and a number of methods have been proposed and demonstrated aimed at suppressing intrachannel nonlinear distortion, through optimization of the dispersion map and pre-compensation [11–16], the use of optimized signal formats [17], in particular the use of phase modulated return-to-zero formats [18–26], alternate-polarization RZ [27–29], subchannel multiplexing [30], polarization mode dispersion-supported transmission [31] and channel precoding [32].

In this paper, we describe experimental and numerical techniques developed to characterize intrachannel cross-phase modulation and four-wave-mixing in long-haul 40 Gbit/s systems. We explain how these techniques can be used to quantify the improvements in performance achievable using new methods, including optimized dispersion maps and novel modulation formats.

2.   CHARACTERIZING INTRACHANNEL NONLINEAR DISTORTION

One challenge faced in characterizing any physical effect in optical signal transmission is that of isolating it from the numerous other effects which occur simultaneously. The goal of our research was to isolate and quantify the impact on the system performance of intrachannel XPM and FWM, separating it from other effects including amplifier noise, chromatic dispersion, polarization mode dispersion and transmitter and receiver patterning.

Methods of measuring the nonlinear distortion include direct BER and Q-factor measurements, and measurements of the signal eye diagrams and signal waveforms using sampling oscilloscopes, for a range of signal launch powers and transmission distances. In the following sections, simulations and experimental measurements of nonlinear pulse interactions in 40 Gbit/s transmission, carried out with the group's recirculating fibre loop testbed, are described.

2.1   Signal waveform distortion

An effective method of characterizing the effects of intrachannel nonlinear distortion is the direct measurement or calculation of the signal waveform distortion after transmission. Three measures of distortion can be used: pulse timing jitter, resulting from XPM, pulse amplitude fluctuations arising from both FWM and XPM, and 'ghost' pulse power, transferred from the pulses to the zero bit slots through FWM. The effectiveness of the use of novel signal formats and optimized dispersion maps in reducing nonlinear pulse interactions can be demonstrated by showing the reduction of these three effects.

2.1.1 Optimization of pre-compensation. In our initial studies, timing jitter and ghost pulse power were calculated for typical system parameters using the split-step Fourier algorithm for a single channel 40 Gbit/s link employing standard single-mode fibre (SMF), with dispersion $D = 17$ ps/(nm·km), fibre nonlinear coefficient $\gamma = 1.2$ (W·km)$^{-1}$ and loss $\alpha = 0.21$ dB/km. Each span comprised 60 km of SMF with dispersion compensating fibre following each span (Fig. 1) [8].

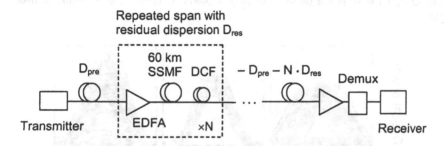

Figure 1. Transmission system with N spans, pre-compensation dispersion D_{pre} and residual dispersion per span D_{res}.

Unchirped RZ pulses, which are more robust than NRZ [17], with 9 ps FWHM were transmitted, encoded with 256-bit random bit sequences. Ideal transmitter and receiver characteristics and the assumption of noiseless amplifiers allowed the investigation of distortion due solely to the fibre nonlinearity. The system considered was a 12 span link, with each span exactly post-compensated by DCF and the optical launch power was 6 dBm. An effective technique to suppress intrachannel nonlinear effects is the optimization of the signal waveform at the input to the link. In particular, the use of simple pre-compensation can have a significant effect on the quality of the received signal [2, 8, 13, 14, 16]. In this study, the effect of using pre-compensation at the transmitter was explored by carrying out multiple simulations with a range of pre-compensation values from 0 to -600 ps/nm. In each case, additional dispersion was added at the receiver to ensure the total link cumulative dispersion was zero. It can be seen from Fig. 2 that the timing jitter varied between values of 0.4 and 1.4 ps, and was minimized at one particular value of pre-compensation, approximately -200 ps/nm.

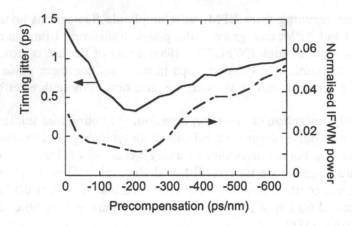

Figure 2. Intrachannel cross-phase modulation (IXPM) timing jitter and intrachannel four-wave-mixing (IFWM) power (normalised to peak power of 'ones') after transmission over 12 spans.

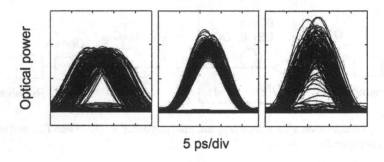

Figure 3. Optical eye diagrams, with D_{pre} = 1 (left) , –240 (centre) and –400 ps/nm (right).

The growth of the four-wave-mixing induced ghost pulse power was also calculated for the same systems. The plot shows that this transmission impairment is also strongly dependent on the input signal waveform, with excessive pre-compensation leading to greater pulse overlap and hence increased FWM. As for the case of timing jitter, the ghost pulse power was minimized with pre-compensation at the transmitter of approximately -200 ps/nm. Fig. 3 shows the eye diagrams of the received signals for three values of precompensation, 0, -240 and -400 ps/nm [8]. With no pre-compensation, timing jitter resulting from XPM dominates, while with excessive pre-compensation of -400 ps/nm, the increased overlap of the pulses results in increased FWM, clearly observable in the eye diagrams as ghost pulses in the zero bit-slots.

2.1.2   Alternate-polarization RZ format transmission.   Following the
simulations characterizing nonlinear pulse interactions in 40 Gbit/s transmis-

sion, recirculating fibre loop experiments were carried out in which the timing
and amplitude jitter, and ghost pulse power, were measured using a high speed
sampling oscilloscope (Fig. 4) [28]. An optical time-division-multiplexing
transmitter was used to obtain the 40 Gbit/s signal, allowing the effects of
varying the relative states of polarization of adjacent pulses to be investigated.

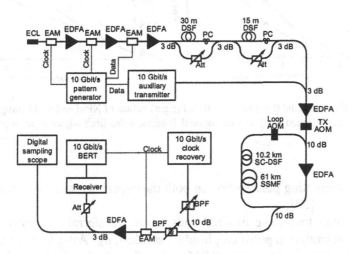

Figure 4. Recirculating loop experimental set-up with optical time-division multiplexed
transmitter (EAM – electroabsorption modulator, SC-DCF – slope compensating dispersion
compensating fibre, PC – polarization controller, AOM – acousto-optic modulator).

Standard single-mode fibre was employed, with the parameters used in the
simulations described in the previous section. The amplifier span length was
61 km and the signal launch power into each span was kept constant at 7 dBm.
Multiple signal traces were measured, and the timing and amplitude jitter were
extracted from the measured waveforms [28]. The measured values of jitter are
plotted in Fig. 5. With parallel polarization states of the pulses, the increase in
the experimentally observed timing jitter due to XPM, σ_t, was 2.4 ps after 7
spans, corresponding to 9.6 % of the bit period. The use of optimal dispersion
pre-compensation of -200 ps/nm at the transmitter in the simulation, not used
in the experiment, led to a lower calculated value of $\sigma_t = 0.9$ ps.

The effect of using alternating pulse polarization states was to reduce σ_t by
32 %, to 1.6 ps after 7 spans, and this was in good agreement with the reduction
predicted by the split-step Fourier simulations. The XPM from adjacent pulses
reduces by 2/3, assuming linear polarization is maintained during transmission,
and reduced by approximately 1/2 if the polarization varies randomly. However, pulses spaced by two bit-slots are still co-polarised, and, as the pulses
spread over multiple bit periods, XPM due to pulses in the same 20 Gbit/s trib-
utary channel is significant, explaining the lower than 2/3 reduction in timing

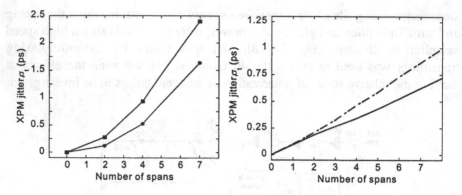

Figure 5. Experimental (left) and calculated (right) values of XPM-induced timing jitter with parallel (squares, dashed line) and orthogonal (circles, solid line) adjacent pulse polarization states.

jitter with alternating polarization in both the experimental measurements and the simulations [28].

Intrachannel four-wave mixing leads to the transferral of energy between the pulses, resulting in pulse amplitude distortion, σ_a. After transmission over 7 spans, the increase in σ_a was 0.046, normalised to the average pulse amplitude (Fig. 6) [28]. The use of alternate polarization states was expected to be effective at reducing the amplitude distortion, as the FWM efficiency reduces to zero for waves with orthogonal polarization states. This was experimentally confirmed, with a reduction in σ_a of 56 % after 7 spans, while the corresponding reduction predicted by the split-step Fourier simulations was 76 %.

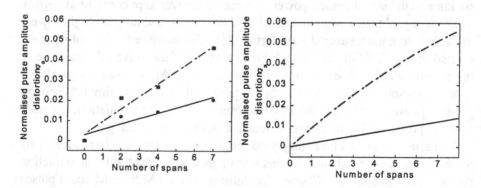

Figure 6. Experimental (left) and calculated (right) values of FWM- and XPM-induced pulse amplitude distortion with parallel (squares, dashed line) and orthogonal (circles, solid line) adjacent pulse polarisation states.

2.1.3 Alternate-phase RZ format transmission.

The work of a number of groups has shown that alternate-phase return-to-zero (AP-RZ) signal format is effective in reducing the distortion caused by nonlinear pulse interactions [18, 20–25]. To investigate this technique, a comprehensive set of experiments were carried out to characterize the effects of intrachannel XPM and FWM through measurements of the signal waveform, taken with a sampling oscilloscope. The AP-RZ transmitter is shown in Fig. 7 [25].

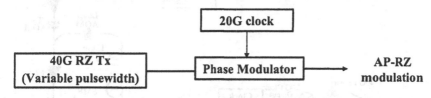

Figure 7. AP-RZ transmitter

Conventional RZ and alternate-phase RZ, with a sinusoidal phase modulation with peak-to-peak amplitude of π rad and a time period twice the bit period, were compared. Initially, the optimum values of pre-compensation were found through simulations, with 5 dBm launch power, which predicted that the optimum amount of pre-compensation is dependent on the format used. The values of -250 ps/nm and -100 ps/nm were optimum for the conventional RZ and AP-RZ formats respectively.

Next, the systems were experimentally tested using the recirculating fibre loop. The experimental set-up, shown in Fig. 8, employed two cascaded electro-absorption modulators (EAMs) driven by the 40 Gbit/s pattern generator (2^{23}-1 PRBS) and a 40 GHz clock signal respectively, to generate an RZ signal with pulse width of approximately 11 ps [25]. A phase modulator driven by a 20 GHz clock signal was then used to impose a π rad optical phase difference between adjacent pulses to generate the AP-RZ format. The signal was launched into the recirculating loop consisting of a span of 60 km SMF post-compensated by 12 km of dispersion compensating fiber. The span residual dispersion of -6 ps/nm, was compensated by a tunable dispersion compensator before the receiver. The 40 Gbit/s signal was transmitted together with five CW WDM channels (Fig. 8) to allow the control of the signal launch power and OSNR (described in more detail in the following section). At the loop output the signal channel was filtered using one port of a flat-top AWG, detected with a photodiode with bandwidth > 45 GHz. Pre- and post- compensation, with optimum values of dispersion determined by the simulations, were experimentally implemented by cascading appropriate lengths of normal and anomalous dispersion fiber, ensuring that the cumulative dispersion over the entire transmission distance was zero. The signal power launched into the pre- and post-compensation was approximately 0 dBm.

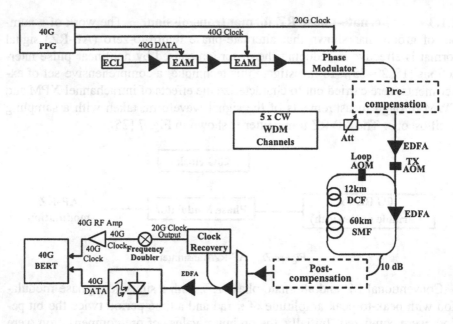

Figure 8. Experimental recirculating loop set-up with AP-RZ transmitter and pre-compensation

The received RZ and AP-RZ signal waveforms, back-to-back and after transmission over 240 km with 10 dBm launch power, with and without pre-compensation, were measured using a high speed sampling oscilloscope.

The waveform of the signals encoding the 11111011111 bit sequence at the input of the link is plotted in Fig. 9(a) [25]. The effect of intrachannel four-wave-mixing can be clearly seen in the waveforms after transmission in Figs. 9(b) and 9(c). With the conventional RZ format, the bit slot carrying the centre zero bit becomes indistinguishable after transmission over 240 km due to the growth of the ghost pulse. Switching on the phase modulator in the transmitter significantly reduces the impact of FWM between the pulses, reducing the growth of the shadow pulse, as can be seen in Fig. 9(c). Figs. 9(d) and 9(e) show the effect of optimizing the pre-compensation, with the zero bit becoming clearly defined in both cases.

Methods of characterizing nonlinear pulse interactions through the measurement and calculation of the distortion of the received signal waveforms have been described in this section. These techniques are useful in understanding the nonlinear effects and demonstrating their suppression through optimized signal formats and dispersion maps. A second method of quantifying the nonlinear pulse interactions is to measure the effect of varying the signal launch power on the received signal Q-factor and bit-error-rate, and in the next section, experiments based on direct BER measurements are described.

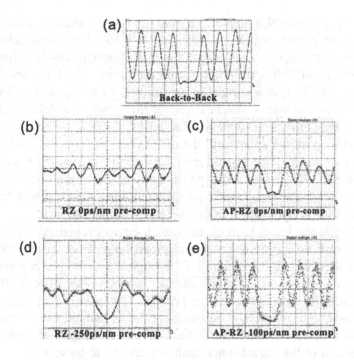

Figure 9. Experimentally measured 11111011111 bit sequence before and after transmission over 240 km of standard single-mode fibre with 10 dBm launch power showing the FWM-induced ghost pulse and its suppression through the use of the AP-RZ signal format and optimized pre-compensation. (a) Output from transmitter. Signals after transmission: (b) RZ with no pre-compensation, (c) AP-RZ with no pre-compensation, (d) RZ with -250 ps/nm pre-compensation and (e) AP-RZ with -100 ps/nm pre-compensation.

2.2 Characterizing nonlinear distortion using BER measurements

In this method, the signal power launched into the spans is varied, and the range of launch powers which achieve a given bit-error rate, for example BER $< 10^{-9}$, after transmission over a given distance is recorded. If the optical signal-to-noise ratio (OSNR) is kept constant while the signal launch power is varied, the measured BER values give a clear indication of the effects of signal distortion due to fibre nonlinearity. Hence, in these measurements, it is important to be able to vary the signal power without varying the noise figure of the loop EDFA [14, 25]. In the experiments described in this section, this was achieved by combining five CW WDM channels with the 40 Gbit/s channel at the input to the recirculating fibre loop (Fig. 8). The signal power was varied by changing the coupling ratio between the 40 Gbit/s channel and the CW channels whilst keeping constant the pump currents of the loop EDFA and therefore its gain, total output power and noise figure [25].

2.2.1   Optimizing the dispersion map.   Fig. 10 shows a measurement of the range of launch powers for which BER < 10^{-9} for a single channel 40 Gbit/s system employing conventional RZ signal format, non-zero dispersion-shifted fibre (NZ-DSF) with D = 4 ps/(nm.km), with a 75 km amplifier span length and span dispersion compensation following each span. The lower values of the launch power represent the noise limit, below which the OSNR at the receiver was too low to achieve BER < 10^{-9}. Increasing the signal launch power into the amplifier spans until the BER increased to 10^{-9} at each distance allowed the tolerance of the signal format to nonlinear effects to be determined. The plot in Fig. 10 shows launch power limits without pre-compensation at the transmitter, and with optimum pre-compensation of -40 ps/nm following the transmitter. In both cases, optimized post-compensation was used at the receiver. While the lower limit on the launch powers remained relatively un-affected, as optimizing the dispersion map had little effect on the OSNR, the upper limit was increased by up to 2 dB through the use of pre-compensation. The optimized dispersion map allowed the distance with BER < 10^{-9} to be increased from 825 km (11 spans) to 1125 km (15 spans). Using this experimental technique of measuring the upper and lower launch power limits while keeping the OSNR constant allowed the cause of the improved performance, the suppression of the intrachannel nonlinear effects, to be verified.

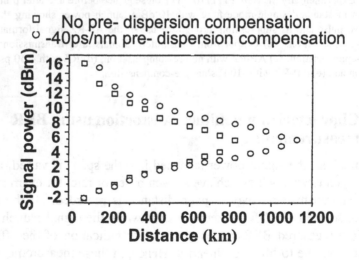

Figure 10.   Measured upper and lower limits on optical launch powers in 40 Gbit/s transmission over non-zero dispersion-shifted fibre to achieve BER < 10^{-9}.

2.2.2   BER measurements with alternate-phase RZ signal format.
The same measurement technique was employed to investigate the improvement in performance in the transmission of the AP-RZ signals described in the

previous section. The measured upper limits to the signal launch power for the 40 Gbit/s RZ and AP-RZ transmission are shown in Fig. 11 [25]. It can be seen that with no pre-compensation, the alternate-phases of adjacent pulses in the AP-RZ format signals resulted in a 2.6 dB increase in the nonlinearity-limited launch power after 240 km compared to that with the conventional RZ format. With pre-compensation of -250 ps/nm, the nonlinear limit for the conventional RZ format was increased by 4.5 dB for the same distance. The highest nonlinear limit was obtained with the AP-RZ format, with -100 ps/nm pre-compensation. In this case, the maximum launch power was 0.5 dB higher than the optimum values with RZ at a distance of 240 km [25].

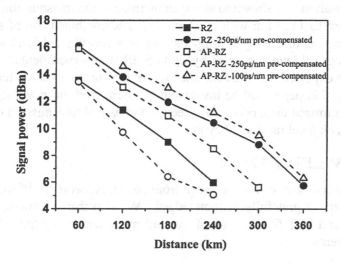

Figure 11. Experimentally measured nonlinearity limited launch powers to achieve BER < 10^{-9} in transmission over standard single-mode fibre with pre-compensation values of 0, -100 and -250 ps/nm.

3. CONCLUSIONS

We have described how intrachannel nonlinear effects lead to distortion in ultra-high bit-rate transmission systems, and discussed the importance of isolating and characterizing these effects to determine optimum design rules for the suppression of pulse interactions. Techniques for characterizing intra-channel XPM and FWM, including the calculation and measurement of the signal waveform distortion and measurements of the received signal BER, were discussed and examples of the use of these characterization methods were presented. Calculations of the pulse timing jitter and ghost pulse powers were used in determining the optimum values of pre-compensation. Measure-ments of timing jitter and pulse amplitude fluctuations, taken with a high-speed sampling oscilloscope, were used in comparison of parallel- and alternate-polarization RZ formats in optical time-division multiplexed transmission. Measurement of the reduced growth of ghost pulses in the zero bit slots in alternate-phase RZ identified the reduced impact of FWM as the explanation for the improvement in the performance of this novel signal format compared to conventional RZ.

Measurements of bit-error-rates for a range of optical launch powers and transmission distances were carried out to characterize nonlinear pulse interactions in systems employing both standard SMF and non-zero dispersion shifted fibre, with a variety of signal formats and dispersion maps. In conventional RZ transmission over non-zero dispersion-shifted fibre, the use of optimum pre-compensation was shown to allow an increase in the transmission distance from 825 km to 1125 km with BER $< 10^{-9}$. The combined use of alternate-phase RZ and optimized pre-compensation was investigated, and increased maximum optical launch powers by up to 5 dB were experimentally demonstrated. It is expected that the continued use of the characterization techniques described in this paper will be invaluable in future work on novel signal formats and optimized dispersion maps required to extend transmission distances at 40 Gbit/s channel rates and beyond.

ACKNOWLEDGMENTS

The financial support for this work from Nortel Networks, EPSRC and the Royal Society is gratefully acknowledged. We also thank Anritsu, Agilent, Yokogawa and SHF for the loan of test and measurement equipment used in the experiments.

REFERENCES

[1] P. V. Mamyshev and N. A. Mamysheva, "Pulse-overlapped dispersion-managed data transmission and intrachannel four-wave-mixing," *Opt. Lett.*, vol. 24, pp. 1454–1456, 1999.

[2] R.-J. Essiambre, B. Mikkelsen, and G. Raybon, "Intrachannel cross phase modulation and four wave mixing in high-speed TDM systems," *Electron. Lett.*, vol. 35, pp. 1576–1578, 1999.

[3] A. Mecozzi, C. B. Clausen, and M. Shtaif, "Analysis of intrachannel nonlinear effects in highly dispersed optical pulse transmission," *IEEE Photon. Tech. Lett.*, vol. 12, no. 4, pp. 392–394, 2000.

[4] M. J. Ablowitz and T. Hirooka, "Resonant nonlinear intrachannel interactions in strongly dispersion-managed transmission systems," *Opt. Lett.*, vol. 25, no. 24, pp. 1750–1752, 2000.

[5] P. Johannisson, D. Anderson, A. Berntson, and J. Mårtensson, "Generation and dynamics of ghost pulses in strongly dispersion-managed fiber-optic communication systems," *Opt. Lett.*, vol. 26, no. 16, pp. 1227–1229, 2001.

[6] F. Neddam and S. Wabnitz, "Role of intra and interchannel cross-phase modulation in higher order fiber dispersion management," *IEEE Photon. Technol. Lett.*, vol. 12, no. 7, pp. 798–800, 2000.

[7] A. Mecozzi, C. B. Clausen, and M. Shtaif, "System impact of intrachannel nonlinear effects in highly dispersed optical pulse transmission," *IEEE Photon. Technol. Lett.*, vol. 12, pp. 1633–1635, 2000.

[8] R. I. Killey, H. J. Thiele, V. Mikhailov, and P. Bayvel, "Reduction of intrachannel nonlinear distortion in 40 Gb/s-based WDM transmission over standard fiber," *IEEE Photon. Techol. Lett.*, vol. 12, no. 12, pp. 1624–1626, 2000.

[9] S. Kumar, "Intrachannel four-wave mixing in dispersion managed RZ systems," *IEEE Photon. Technol. Lett.*, vol. 13, no. 8, pp. 800–802, 2001.

[10] T. Hirooka and M. J. Ablowitz, "Analysis of timing and amplitude jitter due to intrachannel dispersion-managed pulse interactions," *IEEE Photon. Technol. Lett.*, vol. 14, no. 5, pp. 633–635, 2002.

[11] B. Konrad and K. Petermann, "Optimum fiber dispersion in high-speed TDM systems," *IEEE Photon. Technol. Lett.*, vol. 13, no. 4, pp. 299–301, 2001.

[12] A. Pizzinat, A. Schiffini, F. Alberti, F. Matera, A. N. Pinto, and P. Almeida, "40-Gb/s systems on G.652 fibers: comparison between periodic and all-at-the-end dispersion compensation," *J. Lightwave Technol.*, vol. 20, no. 9, pp. 1673–1679, 2002.

[13] Y. Frignac, J.-C. Antona, S. Bigo, and J.-P. Harnaide, "Numerical optimization of pre- and in-line dispersion compensation in dispersion-managed systems at 40 Gbit/s," in *Proc. Conference on Optical Fiber Communication (OFC 2002)*, (Anaheim, California), paper ThFF5, Mar. 2002.

[14] R. I. Killey, V. Mikhailov, S. Appathurai, and P. Bayvel, "Investigation of nonlinear distortion in 40-Gb/s transmission with higher order mode fiber dispersion compensators," *J. Lightwave Technol.*, vol. 20, no. 12, pp. 2282–2289, 2002.

[15] A. Del Duce, R. I. Killey, and P. Bayvel, "Comparison of nonlinear pulse interactions in 160-Gb/s quasi-linear and dispersion managed soliton systems," *J. Lightwave Technol.*, vol. 22, no. 5, pp. 1263–1271, 2004.

[16] H. Xu, J. Zweck, L. Yan, C. R. Menyuk, and G. M. Carter, "Quantitative experimental study of intrachannel nonlinear timing jitter in a 10-Gb/s terrestrial WDM return-to-zero system," *IEEE Photon. Technol. Lett.*, vol. 16, no. 1, pp. 314–316, 2004.

[17] A. Hodžić, B. Konrad, and K. Petermann, "Alternative modulation formats in N × 40 Gb/s WDM standard fiber RZ-transmission systems," *J. Lightwave Technol.*, vol. 20, no. 4, pp. 598–607, 2002.

[18] A. Sano and Y. Miyamoto, "Performance evaluation of prechirped RZ and CS-RZ formats in high-speed transmission systems with dispersion management," *J. Lightwave Technol.*, vol. 19, no. 12, pp. 1864–1871, 2001.

[19] R. Ohhira, D. Ogasahara, and T. Ono, "Novel RZ signal format with alternate chirp for suppression of nonlinear degradation in 40 Gb/s based WDM," in *Proc. Conference on Optical Fiber Communication (OFC 2001)*, (Anaheim, California), paper WM2, Mar. 2001.

[20] J. Martensson, J. Li, A. Berntson, A. Djupsjobacka, and M. Forzati, "Suppression of intrachannel four-wave-mixing by phase modulation at one quarter bit rate,", *Electron. Lett.*, vol. 38, no. 23, pp. 1463–1465, 2002.

[21] M. Forzati, J. Mårtensson, A. Berntson, A. Djupsjobacka, and P. Johanisson, "Reduction of intrachannel four-wave mixing using the alternate-phase RZ modulation format," *IEEE Photon. Technol. Lett.*, vol. 14, no. 9, pp. 1285–1287, 2002.

[22] K. S. Cheng and J. Conradi, "Reduction of pulse-to-pulse interaction using alternative RZ formats in 40-Gb/s systems," *IEEE Photon. Technol. Lett.*, vol. 14, pp. 98–100, 2002.

[23] A. V. Kanaev, G. G. Luther, V. Kovanis, S. R. Brickham, and J. Conradi, "Ghost-pulse generation suppression in phase-modulated 40 Gbit/s RZ transmission," *J. Lightwave Technol.*, vol. 21, pp. 1486–1489, 2003.

[24] D. M. Gill, A. H. Gnauck, X. Liu, X. Wei, and Y. Su, "$\pi/2$ alternate-phase on-off keyed 42.7 Gb/s long haul transmission over 1980 km of standard single-mode fiber," *IEEE Photon. Technol. Lett.*, vol. 16, no. 3, pp. 906–908, 2004.

[25] S. Appathurai, V. Mikhailov, R. I. Killey, and P. Bayvel, "Investigation of the optimum alternate-phase RZ modulation format and its effectiveness in the suppression of intra-channel nonlinear distortion in 40-Gb/s transmission over standard single-mode fiber," *IEEE J. Select. Topics Quantum Electron.*, vol. 10, no. 2, pp. 239–249, 2004.

[26] N. Alić and Y. Fainman, "Data-dependent phase coding for suppression of ghost pulses in optical fibers," *IEEE Photon. Technol. Lett.*, vol. 16, no. 4, pp. 1212–1214, 2004.

[27] F. Matera, M. Settembre, M. Tamburrini, F. Favre, D. Le Guen, T. Georges, M. Henry, G. Michaud, P. Franco, A. Schiffini, M. Romagnoli, M. Guglielmucci, and S. Cascelli, "Field demonstration of 40 Gb/s soliton transmission with alternate polarizations," *J. Lightwave Technol.*, vol. 17, no. 11, pp. 2225–2234, 1999.

[28] V. Mikhailov, R. I. Killey, S. Appathurai, and P. Bayvel, 'Investigation of intrachannel nonlinear distortion in 40 Gbit/s transmission over standard fibre," in *Proc. European Conference on Optical Communication (ECOC 2001)*, (Amsterdam, The Netherlands), paper Mo.L.3.4, Oct. 2001.

[29] X. Chongjin, K. Inuk, A. H. Gnauck, L. Möller, L. F. Mollenauer, and A. R. Grant, "Suppression of intrachannel nonlinear effects with alternate-polarization formats," *J. Lightwave Technol.*, vol. 22, no. 3, pp. 806–812, 2004.

[30] J. Zweck and C. R. Menyuk, "Reduction of intrachannel four-wave-mixing using sub-channel multiplexing," *IEEE Photon. Technol. Lett.*, vol. 15, no. 2, pp. 323–325, 2003.

[31] L. Möller, S. Yikai, G. Raybon, and X. Liu, "Polarization-mode-dispersion-supported transmission in 40 Gb/s longhaul systems," *IEEE Photon. Technol. Lett.*, vol. 15, no. 2, pp. 335–337, 2003.

[32] E. G. Shapiro, M. P. Fedoruk, S. K. Turitsyn, and A. Shafarenko, "Reduction of non-linear intrachannel effects by channel asymmetry in transmission lines with strong bit overlapping," *IEEE Photon. Technol. Lett.*, vol. 15, no. 10, pp. 1473–1475, 2003.

MATHEMATICAL AND EXPERIMENTAL ANALYSIS OF INTERFEROMETRIC CROSSTALK NOISE INCORPORATING CHIRP EFFECT IN DIRECTLY MODULATED SYSTEMS

Efraim Buimovich-Rotem and Dan Sadot
Ben Gurion University, POB 653 Beer-Sheva, 84105 Israel. E-mail: buimov@ee.bgu.ac.il Sadot@ee.bgu.ac.il

Abstract: We derive mathematically and verify experimentally the effect of thermal chirp on interferometric crosstalk reduction in directly modulated systems.

Key words: interferometric crosstalk; thermal chirp.

1. INTRODUCTION

Interferometric homodyne crosstalk is a well known detrimental effect in fiber optic communication networks. The induced power penalty in directly modulated systems is lower as compared to externally modulated systems due to chirp [1–3]. As the frequencies of the desired signal and interferer change due to chirp the mixing noise may have a non-zero intermediate frequency (which equals the difference in frequencies), and therefore is filtered by the receiver. In DFB lasers thermal chirp is the dominant cause for the reduced penalty, while adiabatic chirp and transient chirp have a lesser effect: adiabatic chirp affects the mixing between mark bits and space bits which has a negligible contribution to power penalty, and transient chirp is avoided by proper design of the laser and driving. In this work we present theoretical and experimental analysis of the crosstalk induced power penalty based on mathematical analysis of the PDF (Probability Distribution Function) of the crosstalk noise generated by homodyne mixing of directly modulated DFB lasers.

2.   CROSSTALK PDF CALCULATION

We define the signal field as

$$E_1(t) = \sqrt{p}D_1(t)\cos(\omega_1(t) \cdot t + \theta_1(t)) \tag{1}$$

and the interferer field as

$$E_2(t) = \sqrt{p}D_2(t)\sqrt{x}\cos(\omega_2(t) \cdot t + \theta_2(t)) \tag{2}$$

where p is relative to the signal power, $\omega_1(t)$, $\omega_2(t)$ are the angular frequencies of the fields. These frequencies change due to chirp. $D_1(t)$, $D_2(t)$ represent the digital data stream with 1 for a mark bit and $\sqrt{xr}$ for a space bit, xr being the modulation extinction ratio. $\theta_1(t)$, $\theta_2(t)$ are the random phase noise processes of the lasers. The current created by the mixing of these fields is:

$$I_X(t) = A\cos(\omega_{IF}(t)t + \psi(t)) \tag{3}$$

where $A = pD_1(t)D_2(t)\sqrt{x}$, $\omega_{IF}(t) = \omega_1(t) - \omega_2(t)$ is the beating frequency, and $\psi(t) = \theta_1(t) - \theta_2(t)$. We have assumed a unit responsivity and polarization match and bit timing match between the two fields. We assume an integrate-and-dump receiver and calculate the intensity of the sampled crosstalk noise

$$n(t) = \frac{1}{T}\int_0^T I_X(t)dt = A\,\text{sinc}\left(\frac{\omega_{IF}T}{2\pi}\right)\cos(\psi(t)) \tag{4}$$

where T is the bit period. The conditional PDF of the beat noise is hence given by [ref]

$$g_{n|f}(r, f) = \frac{1}{\pi\sqrt{A^2\,\text{sinc}^2(f_{IF}T) - r^2}} \tag{5}$$

where we use $\omega_{IF} = 2\pi f_{IF}$. In the case of heterodyne mixing between two non-chirped lasers (as in external modulation systems) this equation gives the PDF of the noise as a function of the heterodyne frequency. In the existence of thermal chirp the PDF of the beat noise is given by

$$g_n(r) = \int g_{n|f}(r, f_{IF})g_f(f_{IF})df_{IF} \tag{6}$$

where $g_f(f_{IF})$ is the PDF of the beating frequency, related with the frequency chirp.

Distribution of thermal frequency chirp

In [1, 2], the laser temperature and thermal chirp evolution are calculated as a function of the modulation current waveform. We used the presented method to calculate the PDF of the laser frequency. Mark bits and space bits are considered separately. We observed that for data rates exceeding 100Mb/s the beat frequency distribution (of mark and space bits separately) is well approximated by a truncated Gaussian distribution.

$$g_f(f_{IF}) = \frac{k}{\sqrt{\pi}\sigma_{th}} \exp\left(-\frac{f_{IF}^2}{4\sigma_{th}^2}\right), \qquad 0 < f_{IF} < 4\sigma_{th} \qquad (7)$$

where σ_{th} is the thermal chirp standard variation variance and k is a normalization factor. We assume here mixing between two DFB lasers with similar chirp behaviour. As shown in [2] as the data rate decreases the thermal chirp variance increases which reduces the induced power penalty. The thermal chirp σ_{th} can be found by measurement of the laser thermal time constants [1] and the simulation presented in [3] or alternatively measured directly using the method we presented earlier [4].

The error probability is calculated separately for each of the four combinations of $D_1(t)$, $D_2(t)$ due to the different values of noise amplitude (A in eq. 5), in each case. We assume that mark and space bits have a large frequency difference due to adiabatic chirp and therefore their mixing is filtered completely. The effect of linear crosstalk i.e. the signal due to the detection of $E_2(t)$ is also considered. The total noise PDF is given by convolution with the additive Gaussian noise of the receiver g_{agn}:

$$g_{total} = g_n(r) \otimes g_{agn}(r) \qquad (8)$$

3.   THEORETICAL AND EXPERIMENTAL RESULTS

We measured the standard variation of the thermal chirp of a commercial DFB laser in the method presented in [4]. The results were $\sigma_{th} = 0.12$ GHz/mA and $\sigma_{th} = 0.08$ GHz/mA at 622 Mb/s and 2.5 Gb/s respectively. Based on these measurements we calculated the expected power penalty for homodyne crosstalk when using this laser. The integral in (6) was calculated numerically. In Fig. 1 we present an example of the PDF calculation of (6) that demonstrates the effect of thermal chirp on the PDF. Different modulation currents amplitudes are considered at 2.5 Gb/s. As the current amplitude increases the effect of thermal chirp also increases and the noise PDF migrates from the well known two-pronged shape to a 'centralized' shape.

We then experimentally measured the power penalty. In the experiment the directly modulated DFB laser was mixed with an externally modulated laser rather then performing a delayed self homodyne mixing in order to avoid correlation between the frequency chirp of the signal $E_1(t)$ and interferer $E_2(t)$. The standard variation σ_{th} in (7) was corrected for this case. The extinction ratio was 8dB, and the modulation current was 35 mA p/p. Since the thermal chirp is enhanced by long runs of mark or space bits, therefore sensitive to the

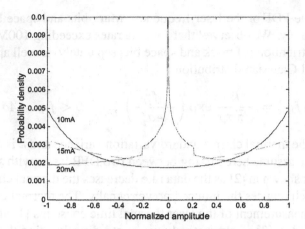

Figure 1. Theoretical PDF of crosstalk noise at 2.5 Gb/s at different modulation currents: blue-10 mA, green-15 mA, red-25 mA.

data pattern, we selected a PRBS $2^{23} - 1$ pattern, rather then 8B10B coding, for example, which is expected to minimize the effect of thermal chirp. Bit alignment and polarization alignment were monitored and maintained.

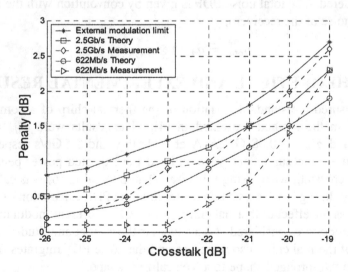

Figure 2. Theoretical and experimental results of homodyne crosstalk induced power penalty.

The theoretical and experimental results are summarized in Fig. 2. A significant power penalty reduction is observed at 622 Mb/s as compared to the external modulation (chirp free) case. This effect appears weaker also at 2.5 Gb/s. The stronger penalty reduction effect at 622 Mb/s is due to both the larger thermal chirp variance and the longer integration time (narrower filtering) at

622 Mb/s. The measured penalties are in good agreement with the theoretical prediction, even though about ∼0.5 dB lower than the theoretical calculations due to imperfect polarization and wavelength match. A modest (almost negligible) improvement in power penalty relative to external modulation is expected at 10 Gb/s systems.

REFERENCES

[1] H. Shalom, A. Zadok, M. Tur, P. J. Legg, W. D. Cornwell, I. Andonovic, "On the various time constants of wavelength changes of a DFB laser under direct modulation," *IEEE J. of Quantum Electronics*, vol. 34, no. 10, pp. 1816–1824, 1998.

[2] W. D. Cornwell, I. Andonovic, A. Zadok, M. Tur, "The role of thermal chirp in reducing interferometric noise in fiber-optic systems driven by directly modulated DFB lasers," *J. Lightwave Technol.*, vol. 18, no. 2, pp. 154–160, 2000.

[3] P. J. Legg, M. Tur, H. Regev, W. D. Cornwell, M. Shabeer, I. Andonovic, "Interferometric noise reduction through intrabit frequency evolution of directly modulated DFB lasers," *J. Lightwave Technol.*, vol. 14, no. 10, pp. 2117–2125, 1996.

[4] E. Buimovich, D. Sadot, "Measurements of thermal frequency chirp in directly modulated DFB lasers," in *Proceedings of LEOS annual meeting 2003*, Tucson AZ, paper TuD7, pp. 208–209.

ON THE IMPACT OF MPI IN ALL-RAMAN DISPERSION-COMPENSATED IMDD AND DPSK LINKS

Stefan Tenenbaum and Pierluigi Poggiolini
PhotonLab, Dipartimento di Eletronica, Politecnico di Torino, Corso Duca Abruzzi 24, 10129 Torino, Italy, stefan.tenenbaum@polito.it, pierluigi.poggiolini@polito.it

Abstract: The impact of MPI in all-Raman multi-span dispersion-compensated links was studied. D+/D- and D+/D-/D+ compensation schemes were considered, for both IMDD and DPSK. OSNR penalties were calculated for several system configurations. Our analysis confirmed that the D+/D-/D+ OSNR penalty is typically much less than that of the D+/D- scheme. We then estimated the increase, due to MPI, in the number of spans required to satisfy a target OSNR for a given total link length, taking into account Kerr nonlinearities. Somewhat unexpectedly, it turned out that such increase can be very significant (up to 15-20%) with the D+/D- scheme and lower but non-negligible (5-10%) with the D+/D-/D+ schemes. DPSK is typically less impacted than IMDD across all configurations.

Key words: Raman amplification; differential phase-shift keying (DPSK); multipath interference (MPI); double Rayleigh back scattering (DRBS); ultra-long-haul.

1. INTRODUCTION

Raman amplifiers can provide improved system OSNR with respect to ED-FAs. However, if the span length is increased to reduce the number of repeaters, the needed Raman gain goes up and MPI (Multi-Path Interference) may develop.

MPI is signal light that reaches the end of the fiber after suffering a double reflection due to Rayleigh Back-Scattering [1]. Although the amount of double-back-scattered light is very low, a high level of Raman gain can make it non-negligible. MPI light is no longer correlated to the originating signal and behaves as noise [2], effectively reducing the OSNR at the RX.

Experimental investigations have shown that effective area management can reduce MPI by properly selecting fiber types and by symmetrically compensating for dispersion in a D+/D-/D+ sequence, as opposed to the conventional D+/D- configuration [3,4].

In this paper we carry out a quantitative theoretical analysis of the impact of MPI generated along multi-fiber-type spans. We compute the OSNR penalty due to MPI for a wide range of span-lengths and total lengths. We investigate both D+/D- and D+/D-/D+ compensation schemes, for both IMDD and DPSK. The penalty plots are both signal-bandwidth and bit-rate independent, so they need not be rescaled if these quantities vary.

Results confirm the great effectiveness of the D+/D-/D+ [4] configuration, which makes the OSNR penalty small in most practical cases. On the contrary, MPI may have a significant impact in D+/D- configurations, both with SMF/DCF and NZDSF/DCF. We also show that DPSK is substantially less sensitive to MPI than IMDD.

We then evaluate the maximum link length vs. span length in the presence and absence of MPI. To obtain meaningful results, we also take into account the level of generated Kerr non-linearity, which we do by constraining the non-linear phase shift k_{NL} [4–7] to a fixed value. This investigation reveals that in certain system configurations, the impact of MPI can be very significant even for the D+/D-/D+ compensation scheme, when such impact is expressed as an increase in number of spans (or a reduction of the span length) caused by MPI for a given total link length.

2.   THEORETICAL APPROACH

Throughout the paper we assume that Raman gain exactly compensates for loss in each span. The MPI power reaching the end of a single-span Raman amplifier made up of just one fiber type can be written as [1]:

$$P_{MPI} = \Gamma_{MPI} \cdot P_{SIG} \tag{1}$$

where P_{SIG} is the average signal power at the end of the fiber and Γ_{MPI} is the MPI coefficient, given by:

$$\Gamma_{MPI} = (R^2 K)/2 \tag{2}$$

with R being the fiber Rayleigh backscattering coefficient, and

$$K = \int_0^L dz_2 \int_0^{z_2} dz_1 \left[e^{-2\alpha_s(z_2 - z_1)} \cdot G_{RA}^2(z_1, z_2) \right], \tag{3}$$

where $G_{RA}(z_1, z_2)$ is the Raman gain in the fiber section between z_1 and z_2 and α_s is signal loss [2]. In compensated spans, more than one fiber type is present and (1) has to be modified to include the contributions of all spliced sections. In D+/D- spans, three MPI contributions must be taken into account: when reflections take place both inside the D+ fiber; both in the D- fiber; one in the D- fiber and one in the D+. In D+/D-/D+ configurations, six contributions must be considered: three from each fiber alone, two from adjacent fibers and another one when reflections occur in the first and last D+ fibers. All these different contributions are depicted in Fig. 1.

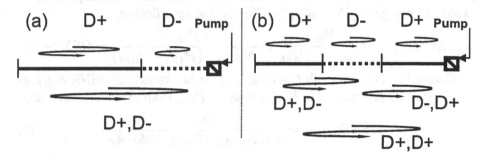

Figure 1.   All the possible dual-reflection contributions to MPI. Case (a) refers to a D+/D- span. Case (b) refers to a D+/D-/D+ span. The various contributions have been labelled according to the fiber types where the reflections occur.

Analytically, Γ_{MPI} for the D+/D- span becomes:

$$\Gamma_{MPI} = R_{D+}^2 K_{D+} + R_{D-}^2 K_{D-} + R_{D+} R_{D-} K_{D+,D-} \qquad (4)$$

and for the D+/D-/D+ span:

$$\Gamma_{MPI} = R_{D+}^2 K_{D+1} + R_{D-}^2 K_{D-} + R_{D+}^2 K_{D+2}$$
$$+ R_{D+} R_{D-} K_{D+,D-} + R_{D-} R_{D+} K_{D-,D+} + R_{D+}^2 K_{D+1,D+2} \qquad (5)$$

where R_{D+} and R_{D-} are the individual fiber backscattering coefficients. The various K's appearing in (5) were derived adapting (3). The full, lengthy expressions of all the K's will be reported elsewhere due to lack of space in this paper. We neglected higher-order (more than two) reflections because they typically become relevant when first-order reflections have already reached system-intolerable levels. Once we have Γ_{MPI}, we can calculate the system OSNR at the RX for a link with N_{SPAN} identical spans:

$$OSNR = \frac{P_{SIG}}{N_{SPAN} \cdot (P_{ASE} + P_{MPI})} \qquad (6)$$

where P_{ASE} and P_{MPI} are the ASE noise and MPI power produced by each Raman amplifier. Note that both the Raman ASE noise and Raman gain were computed taking accurately into account the presence of different fiber types in the span.

Typically, system designers set a target system OSNR to ensure proper operation and margins. We call it $OSNR_T$ and, for an actual system to operate properly, it must be that OSNR from (6) satisfies: $OSNR \geq OSNR_T$. We call this "the $OSNR_T$ constraint". From (1) and (6), the minimum launched signal power required to satisfy $OSNR = OSNR_T$ is given by:

$$P_{SIG} = \frac{N_{SPAN} \cdot OSNR_T \cdot P_{ASE}}{1 - N_{SPAN} \cdot \Gamma_{MPI} \cdot OSNR_T} \qquad (7)$$

If we now used P_{SIG} from (7) in a system where MPI was not present, we would clearly get $OSNR_{noMPI} > OSNR_T$ and specifically:

$$OSNR_{noMPI} = \frac{P_{SIG}}{N_{SPAN} \cdot P_{ASE}} > \frac{P_{SIG}}{N_{SPAN} \cdot (P_{ASE} + P_{MPI})} = OSNR_T \qquad (8)$$

Comparing $OSNR_T$ with $OSNR_{noMPI}$, we can assess the OSNR degradation due to the presence of MPI. Substituting P_{SIG} from (7) into (8) we find:

$$OSNR_{noMPI} = \frac{OSNR_T}{1 - N_{SPAN} \cdot \Gamma_{MPI} \cdot OSNR_T} \qquad (9)$$

and finally the penalty in dB, that we call $\Delta OSNR_{dB}$, is:

$$\begin{aligned} \Delta OSNR_{dB} &= OSNR_{noMPI,dB} - OSNR_{T,dB} \\ &= -10\log_{10}\left(1 - N_{SPAN} \cdot OSNR_T \cdot \Gamma_{MPI}\right) \qquad (10) \end{aligned}$$

Table 1. Fiber Parameters

TABLE I	SLA	SMF	NZDSF	IDF	DCF
α_s [dB/km]	0.19	0.2	0.2	0.25	0.5
α_p [dB/km]	0.24	0.3	0.3	0.31	0.6
D [ps/nm]	+20	+17	+5	-40	-100
A_{eff} [μm^2]	110	80	55	30	25
C_R[1/W/km]	0.29	0.4	0.5	1.2	1.8
R [dB/km]	-44	-42.3	-40	-38	-33.8

3.   PENALTY RESULTS

To analyze the impact of MPI, penalty contour plots of $\Delta OSNR_{dB}$ were drawn for D+/D- and D+/D-/D+ configurations. Different types of D+ fiber were used: SLA, SMF and NZDSF. DCF and IDF were used as D- fibers. Fiber data is shown in Table 1. We computed the OSNR penalty assuming $OSNR_T$ equal to either 14 dB or 11.5 dB. These are reasonable values for IMDD and DPSK, respectively, assuming BER $= 10^{-11}$, the use of FECs and a system margin of 3 to 4 dB. Note that to express all OSNRs, we chose an ASE noise bandwidth equal to the bit rate. This way, all penalty results turn out to be completely independent of the bit rate, i.e., the penalty contour plots apply to any bit rate. Note that, in order to assign individual lengths to the different fiber types within a span, we assumed that dispersion was fully compensated for in each span. We must also point out that certain combinations of

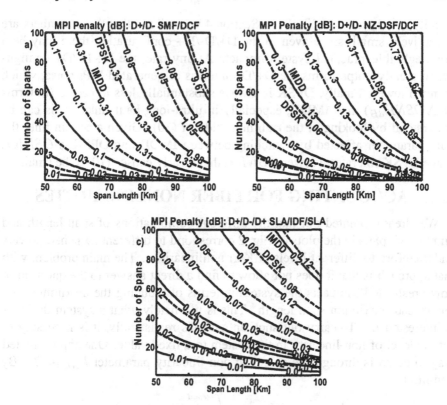

Figure 2. Contour plot of the OSNR penalty in dB, $\Delta OSNR_{dB}$, due to MPI as a function of span length and number of spans for (a) D+/D- with SMF/DCF; (b) D+/D- with NZDSF/DCF; (c) D+/D-/D+ with SLA/IDF/SLA.

span lengths and number of spans in the plots of Fig. 2 may require values of P_{SIG}, from (7), which could excite excessive SPM/XPM/FWM. However, different transmission formats, different values of channel spacing and different system parameters may all lead to very different maximum tolerable launched powers P_{SIG}. Therefore, we decided to first quantify the MPI impact alone and disregard all other impairments. Later, in Section 4, we will try and insert Kerr nonlinearities in the picture, using a simplified approach.

From the penalty contour plots we see that MPI can have a very substantial impact in the D+/D- SMF/DCF configuration. The impact is somewhat lessened by using NZDSF/DCF, thanks to the lower amount of DCF needed to compensate for the smaller NZDSF dispersion, but it may still be significant in long-haul systems.

The D+/D-/D+ SLA/IDF/SLA configuration, using state-of-the-art fibers, shows instead a very low MPI impact all the way up to trans-oceanic lengths, even using very long spans. At 9,000 km, with 90 km spans, the expected penalty is only about 0.22 dB. In Section 4 we will see that these numbers are deceptively small, since even in the D+/D-/D+ configuration there may be a non-negligible practical system impact. At any rate, the D+/D-/D+ configuration greatly outperforms the D+/D- in terms of generated MPI noise, which is much lower. Finally, DPSK is always substantially less impacted (in terms of $\Delta OSNR_{dB}$) than IMDD, especially in ultra-long-haul links. This can be understood by looking at the rightmost side of (10). Given a certain link, the only parameter affected by switching between IMDD and DPSK is $OSNR_T$. Since DPSK tolerates a lower $OSNR_T$, this clearly causes (10) to be smaller.

4.   ACCOUNTING FOR FIBER NONLINEARITIES

We already pointed out that the different combinations of span length and number of spans in the plots of Fig. 2 correspond to different launched powers and therefore to different levels of Kerr nonlinearities. The main problem with that approach is that it does not allow to find a direct answer to the question of how greatly MPI impacts the system in terms of reducing the maximum total length and maximum span length. This is ultimately what a system designer is interested in. To carry out such an analysis realistically, it is necessary to set the level of non-linearity in the system to a fixed value. One approximated way to do so is through the so-called non-linearity parameter k_{NL} [4–7]. By definition:

$$k_{NL} = \gamma \cdot N_{SPAN} \cdot \int_0^{L_{span}} P_{SIG}(z) \cdot dz$$

where $P_{SIG}(z)$ is the signal power (per channel) along the span, γ is the fiber non-linearity coefficient and N_{SPAN} is the number of spans.

k_{NL} has the physical meaning of the total SPM phase shift accumulated by a single channel along the whole system and is measured in radians. Therefore, it is reasonable to think that it somehow "measures" SPM rather accurately. As for the other Kerr effects, the XPM phase shift is directly proportional to k_{NL} whereas FWM is proportional to k_{NL}^2. However, even forcing a constant k_{NL} across different system span lengths and total lengths does not really ensure that we are operating with a constant penalty from non-linearity, because many system parameters such as dispersion maps greatly influence the impact of nonlinearities. However, fixing k_{NL} is the best that can be done in a relatively simple way to account for Kerr effects, so we decided to adopt this method nonetheless.

In practical systems, k_{NL} may vary quite substantially. For instance [6] is a 10,000 km low-non-linearity system where $k_{NL} = 0.7$. On the other hand, in [7] a 9200 km system is reported having a "high" $k_{NL} = 2\pi$. We assumed a total per-channel k_{NL} equal to 1.5 radians, which is a "medium" level of non-linearity. Once a fixed k_{NL} is forced, not all points in the plots

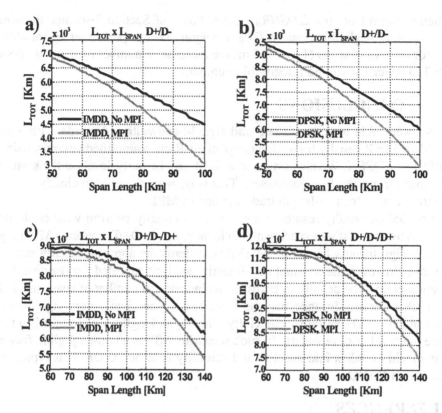

Figure 3. Maximum total system length versus span length with D+/D- SMF/DCF for (a) IMDD and (b) DPSK; with D+/D-/D+ SLA/IDF/SLA for (c) IMDD and (d) DPSK, assuming a constant value of the system non-linearity parameter $k_{NL} = 1.5$.

of Fig. 2 become feasible. Forcing this constraint, it is possible to extract from Fig. 2 those combinations of N_{SPAN} and L_{SPAN} for which $k_{NL} = 1.5$. These combinations will give us the maximum system length as a function of L_{SPAN}. In Fig. 3 we show the results for IMDD. As expected the impact of MPI is substantial in the D+/D- SMF/DCF case. For instance, in the absence of MPI it is possible to use 100-km spans to reach 4500 km using IMDD. However, the presence of MPI forces a reduction of the span length to 86 km, which translates into a 16% increase in the number of spans needed. This is a rather large impact in terms of system economics. With DPSK, there is a 13.5% increase at 6000 km.

Surprisingly, the impact of MPI may also be substantial in the D+/D-/D+ SLA/IDF/SLA case. If we pick IMDD with the typical trans-pacific length of 8400 km, we see that in the absence of MPI it would theoretically be possible to use 100 km spans. However, when MPI is present, L_{SPAN} goes down to 91 km. N_{SPAN}, as a result goes up 10%, a large number for a D+/D-/D+

scheme which from the $\Delta OSNR_{dB}$ calculations of Section 3 seemed to incur a completely negligible penalty. DPSK is again a better performer and at 8500 km for ultra-long spans (130-140 km) the increase in number of spans is about 5.5%, a lower but still non-negligible number.

5.   CONCLUSIONS

We were able to draw bit-rate and signal-bandwidth-independent plots of the MPI OSNR penalty for a wide range of all-Raman amplified systems, using IMDD and DPSK. We also imposed a fixed non-linearity level on the system by constraining the k_{NL} parameter. This way, we were able to clearly assess the total-length to span-length tradeoffs due to MPI.

The OSNR penalty results turned out to essentially confirm what could be inferred by the existing experimental evidence, i.e., in D+/D- systems MPI may be very significant whereas in D+/D-/D+ systems the apparent penalty values are low. However, the constant non-linearity analysis revealed a somewhat different picture. The penalty in terms of the increase in number of spans needed to reach a certain distance may be large and economically very significant, even in D+/D-/D+ systems, especially when span lengths are stretched to reduce the number of repeaters. In such scenarios MPI is a non-negligible factor that should be taken into account and carefully dealt with. DPSK is typically better than IMDD.

REFERENCES

[1] C. R. S. Fludger, and R. J. Mears, "Electrical measurements of multipath interference in distribute Raman amplifiers," *J. Lightwave Technol.*, vol. 19, pp. 536–545, 2001.

[2] V. Curri, and G. Rizzo, "Statistical properties and system impact of multipath interference in Raman amplifiers," in *Proc. European Conference on Optical Communication (ECOC' 2001)*, (Amsterdam, Holland), pp. 110–111, Sept. 2001.

[3] M. Vasilyev *et al.*, in *Proc. Optical Fiber Communication Conference and Exhibit (OFC' 2002)*, (Anaheim-USA), pp. 508–509, Mar. 2002.

[4] R. Hainberger *et al.*, "Comparison of span configurations of Raman-amplified dispersion-managed," *IEEE Photon. Technol. Lett.*, vol. 14, pp. 471–473, 2002.

[5] V. Curri, "System advantages of Raman amplifiers," in *Proc. National Fiber Optic Engineers Conference (NFOEC' 2000)*, (San Diego, USA), vol. 1, pp. 35–46, 2000.

[6] C. Rasmussen *et al.*, "DWDM 40 G Transmission over trans-pacific distance (10000 km) using CSRZ-DPSK, enhanced FEC, and all-Raman-amplified 100-km UltraWave fiber spans," *IEEE Photon. Technol. Lett.*, vol. 14, pp. 203–207, 2004.

[7] T. Mizuochi *et al.*, "A comparative study of DPSK and OOK WDM transmission over transoceanic distances and their performance degradations due to nonlinear phase noise," *J. Lightwave Technol.*, vol. 21, pp. 1933–1943, 2003.

IV

MODULATION FORMATS AND DETECTION

MODULATION FORMATS FOR OPTICAL FIBER TRANSMISSION
Invited Paper

Klaus Petermann
Technische Universität Berlin, Fachgebiet Hochfrequenztechnik, D-10587 Berlin, Germany
petermann@tu-berlin.de

Abstract: Modulation formats using the signal space in the amplitude, phase and polarization domain are reviewed.

Key words: modulation formats; optical fiber transmission; return-to-zero; nonreturn-to-zero; differential phase-shift keying; differential quadrature phase-shift keying.

1.   INTRODUCTION

The most simple method to transmit digital data consists of a simple on-off keying technique. However, as simply illustrated in Fig. 1, for advanced modulation formats we can not only make use of the light intensity, but we can also make use of its phase and polarization.

The goal for designing novel modulation formats consists of achieving noise-tolerant and distortion-tolerant transmission systems. For receiving these modulated signals both coherent and incoherent receivers may be used. In this overview we will restrict ourselves to modulation formats where direct (incoherent) detection can be used.

2.   CLASSIFICATION OF MODULATION FORMATS

As a first classification we may distinguish between phase and amplitude modulated signals. But even if we just consider modulation formats based on amplitude keying there is a great variety as outlined in Fig. 2 [1].

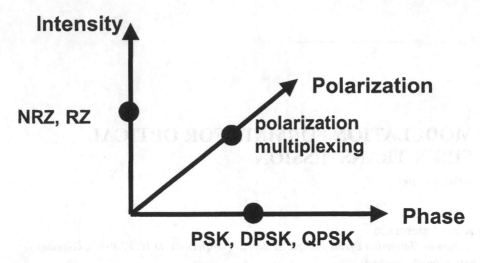

Figure 1.   Signal space in intensity, phase and polarization for novel modulation formats.

The simplest method consists of simple binary on-off keying where we can either use NRZ (non-return-to-zero) or RZ (return-to-zero) formats which may be combined also with chirp [2], yielding, e.g. single-sideband based (VSB, SSB) formats [3], chirped NRZ (C-NRZ) [4] or chirped RZ (CRZ, ACRZ). Phase coding may result in CSRZ (carrier-suppressed RZ) [5], duobinary (DB) [6] or alternate-mark-inversion (AMI) [7] formats.

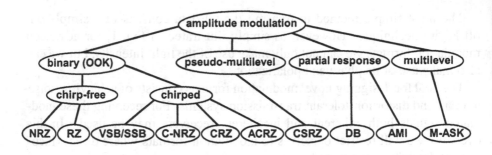

Figure 2.   Amplitude modulation formats [1].

As an alternative to amplitude modulation formats there is a growing interest in phase modulation formats as illustrated in Fig. 3 [1]. In particular, by using binary RZ-DPSK (DPSK-differential phase shift keying) record transmission distances could be achieved [8]. Phase modulation techniques are attractive also for multi-level modulation, where in particular quaternary DPSK (DQPSK) is of special interest [9].

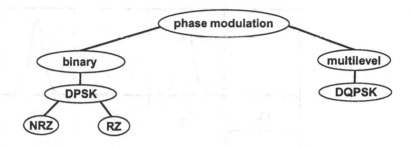

Figure 3. Phase modulation formats [1].

It is very difficult to make general statements on the performance difference between different modulation formats. However, a few general guidelines can be given:

RZ behaves better than NRZ with respect to nonlinear impairments in the fiber line [10] and with respect to receiver sensitivity. RZ suffers, however, from a stronger signal degradation due to chromatic dispersion.

Duobinary and multi-level formats exhibit a lower spectral width and thus allow for a higher spectral efficiency. They are also more tolerant with respect to chromatic dispersion.

By using balanced receivers phase modulated signals achieve a higher receiver sensitivity than similar amplitude modulated signals. E.g., binary DPSK balanced receivers exhibit a 3dB sensitivity improvement over on/off-keying [11].

3. EXAMPLES

In order to demonstrate that the actual performance prediction is not trivial an RZ transmission with very tight optical filtering (corresponding to 40 Gbit/s channels with 50 GHz channel spacing) are considered. If we consider there in Fig. 4 in an RZ-based amplitude modulated signal a sequence of 4 succeeding pulses with different phases, we may attribute the phase signatures to different modulation formats [12] including alternate polarization schemes (denoted as APol in Fig. 4) if the polarization is switched from pulse to pulse.

Even though the differences between these modulation formats in Fig. 4 appear small, there are substantial differences between the eye diagrams of these formats if tight optical filtering is applied (as required for 40 Gbit/s systems with 50 GHz channel spacing) as shown in Fig. 5 [12]. Without chromatic dispersion there is an advantage for the AP90-RZ scheme (90° phase shift between pulses), but for increasing chromatic dispersion the best results are achieved for the APol-CS-RZ scheme (alternate polarization carrier suppressed RZ). Fiber nonlinearities (IFWM) can be reduced by AP90-RZ and

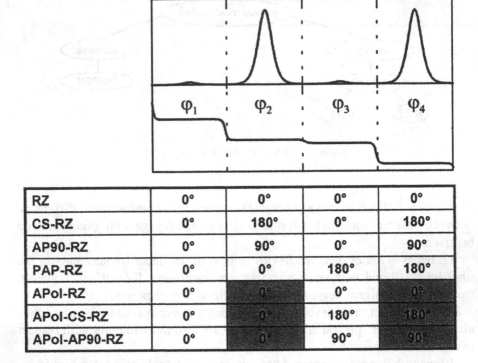

	φ_1	φ_2	φ_3	φ_4
RZ	0°	0°	0°	0°
CS-RZ	0°	180°	0°	180°
AP90-RZ	0°	90°	0°	90°
PAP-RZ	0°	0°	180°	180°
APol-RZ	0°	0°	0°	0°
APol-CS-RZ	0°	0°	180°	180°
APol-AP90-RZ	0°	0°	90°	90°

Figure 4.   On/off keyed RZ signals with different phase and polarization signatures [12].

PAP-RZ [13, 14]. Polarization switched schemes are also exhibiting improved performance for NRZ-based schemes [15].

By combining polarization multiplexing and multi-level modulation schemes very high spectral efficiencies can be obtained, with a spectral efficiency of, e.g. 1.6 bit/s/Hz for a polarization multiplexed DQPSK scheme [16, 17].

4.   SUMMARY AND CONCLUSIONS

One must be very careful with general statements for pro and con with respect to specific modulation formats, but a few guidelines have been identified as follows:

- RZ behaves better than NRZ, but it occupies a larger spectral width.

- Duobinary and multi-level formats allow for a higher spectral efficiency and higher chromatic dispersion.

- DPSK has a 3dB sensitivity advantage if balanced receivers are used.

- Transmitting the data in alternate polarizations may also yield a performance improvement.

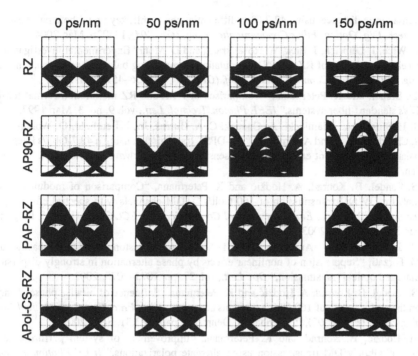

Figure 5. Dispersion Tolerance of on/off keyed RZ signals with different phase signatures in the case of 50 GHz optical filtering [12].

REFERENCES

[1] P. J. Winzer, R.-J. Essiambre, "Advanced optical modulation formats," in *Proc. Eur. Conf. Optical Communication (ECOC'03)*, (Rimini, Italy), paper Th.2.6.1, Sept. 2003.

[2] A. Hodžić, B. Konrad, and K. Petermann, "Alternative modulation formats in N x 40 Gb/s WDM standard fiber RZ-transmission systems," *J. Lightwave Technol.*, vol. 20, no. 4, Apr. 2002.

[3] S. Bigo, "Improving spectral efficiency by ultra-narrow optical filtering to achieve multiterabit/s capacity", in *Tech. Dig. Optical Fiber Communication Conf. (OFC'01)*, pp. 362–364, Mar. 2001.

[4] A. Hodžić, B. Konrad, and K. Petermann, "Prechirp in NRZ-Based 40-Gb/s single-channel and WDM transmission systems," *IEEE Photon. Technol. Lett.*, vol. 14, no. 2, Feb. 2002.

[5] A. Sano, Y. Miyamoto, "Performance evaluation of prechirped RZ and CS-RZ formats in high-speed transmission systems with dispersion management," *J. Lightwave Technol.*, vol. 19, no. 12, Dec. 2001.

[6] T. Ono, Y. Yano, K. Fukuchi, T. Ito, H. Yamazaki, M. Yamaguchi, and K. Emura, "Characteristics of optical duobinary signals in Terabit/s capacity high-spectral efficiency WDM systems," *J. Lightwave Technol.*, vol. 16, no. 5, May 1998.

[7] P. J. Winzer, A. H. Gnauck, G. Raybon, S. Chandrasekhar, Y. Su, and J. Leuthold, "40-Gb/s return-to-zero alternate-mark-inversion (RZ-AMI) transmission over 2000 km," *IEEE Photon. Technol. Lett.*, vol. 15, no. 5, May 2003.

[8] G. Charlet, E. Corbel, J. Lazaro, A. Klekamp, R. Dischler, P. Tran, W. Idler, H. Mardoyan, A. Konczykowska, F. Jorge, and S. Bigo, "WDM transmission at 6 Tbit/s capacity over

transatlantic distance, using 42.7 Gb/s differential phase-shift keying without pulse carver," in *Tech. Dig. Optical Fiber Communication Conf. (OFC'04)*, PDP36, Mar. 2004.

[9] C. Wree, J. Leibrich, J. Eick, W. Rosenkranz, and D. Mohr,"Experimental investigation of receiver sensitivity of RZ-DQPSK modulation format using balanced detection," in *Tech. Dig. Optical Fiber Communication Conf. (OFC'03)*, pp. 456–457, Mar. 2001.

[10] D. Breuer, and K. Petermann, "Comparison of NRZ and RZ modulation format for 40-Gb/s standard-fiber systems," *IEEE Photon. Technol. Lett.*, vol. 9, no. 3, Mar. 1997.

[11] J. H. Sinsky, A. Adamiecki, A. Gnauck, C. A. Burrus, Jr., J. Leuthold, O. Wohlgemuth, S. Chandrasekhar, and A. Umbach, "RZ-DPSK transmission using a 42.7-Gbit/s integrated balanced optical front end with record sensitivity," *J. Lightwave Technol.*, vol. 22, no. 1, Jan. 2004.

[12] S. Randel, B. Konrad, A. Hodžić and K. Petermann, "Comparison of modulation formats for DWDM transmission of 160 Gbit/s OTDM-channels with spectral efficiency of 0.8 bit/s/Hz," in *Proc. Eur. Conf. Optical Communication (ECOC'03)*, (Rimini, Italy), paper Mo4.2.6, Sept. 2003.

[13] P. Johannisson, A. Anderson, M. Marklund, A. Bernston, A. Djupsjobacka, and M. Forzati, "Suppression of nonlinear effects by phase alternation in strongly dispersion-managed optical transmission," *Opt. Lett.*, vol. 27, pp. 1073–1075, 2002.

[14] S. Randel, B. Konrad, A. Hodžić and K. Petermann, "Influence of bitwise phase changes on the performance of 160 Gbit/s transmission systems," in *Proc. Eur. Conf. Optical Communication (ECOC'02)*, (Copenhagen, Denmark), paper P3.31, Sept. 2002.

[15] A. Hodžić, B. Konrad, and K. Petermann, "Improvement of system performance in N x 40-Gb/s WDM transmission using alternate polarizations," *IEEE Photon. Technol. Lett.*, vol. 15, no. 1, Jan. 2003.

[16] C. Wree, N. Hecker-Denschlag, E. Gottwald, P. Krummrich, J. Leibrich, E.-D. Schmidt, B. Lankl, and W. Rosenkranz, "High spectral efficiency 1.6-b/s/Hz transmission (8 x 40 Gb/s with a 25-GHz grid) over 200-km SSMF using RZ-DQPSK and polarization multiplexing," *IEEE Photon. Technol. Lett.*, vol. 15, no. 9, Sep. 2003.

[17] Y. Zhu, K. Cordina, N. Jolley, R. Feced, H. Kee, R. Rickard, and A. Hadjifotiou, "1.6 bit/s/Hz orthogonally polarized DSRZ-DQPSK transmission of 8x40 Gbit/s over 320 km NDSF," in *Tech. Dig. Optical Fiber Communication Conf. (OFC'04)*, TuF-1, March 2004.

DISPERSION LIMITATIONS IN OPTICAL SYSTEMS USING OFFSET DPSK MODULATION

Jin Wang[1] and Joseph M. Kahn[2]

[1]*Department of Electrical Engineering and Computer Sciences, University of California, Berkeley, CA 94720. (e-mail: wangjin@eecs.berkeley.edu).* [2]*Department of Electrical Engineering, Stanford University, Stanford, CA 94305. (e-mail: jmk@ee.stanford.edu).*

Abstract: In M-ary differential phase-shift keying (DPSK), $\log_2 M$ information bits are encoded in the phase difference $\Delta\phi$ between successive symbols, where $\Delta\phi$ assumes values in the set $\theta + 2(i-1)\pi/M$, $i = 1, \ldots M$. When $\theta = 0$, we have conventional DPSK, while when $\theta \neq 0$, we have offset DPSK. Using numerical analysis, we investigate the performance of binary and quaternary offset DPSK for arbitrary θ in the presence of fiber chromatic dispersion (CD) and first-order polarization-mode dispersion (PMD). We show that the choice of θ may strongly affect the tolerance of offset DPSK systems to CD and PMD.

Key words: optical fiber communication; chromatic dispersion; polarization-mode dispersion; differential phase-shift keying (DPSK); modulation formats.

1. INTRODUCTION

Differential phase-shift-keying (DPSK) is a class of modulation techniques that encodes $\log_2 M$ information bits in the phase difference $\Delta\phi$ between successive symbols, with $\Delta\phi$ assuming values in the angle set $\Omega = \{\theta + 2\pi(i - 1)/M, \ i = 1, \ldots M\}$. When $\theta = 0$, we have conventional DPSK, while when $\theta \neq 0$, we have offset DPSK [1, 2]. The parameter M is the dimension of the signal set, with $M = 2$ and 4 corresponding to binary and quaternary DPSK, respectively. In this paper, to denote specific values of θ and M, we use the notation θ-M-DPSK. When θ is indefinite, we use M-DPSK.

In the last several years, 2-DPSK and 4-DPSK have been extensively employed in high-capacity, long-haul optical transmission experiments. Initial experiments used only conventional DPSK [3,4], i.e., 0-2-DPSK and 0-4-DPSK, in part, because of the ease of generating $\{0, \pi\}$ or $\{0, \pi/2, \pi, 3\pi/2\}$ phase

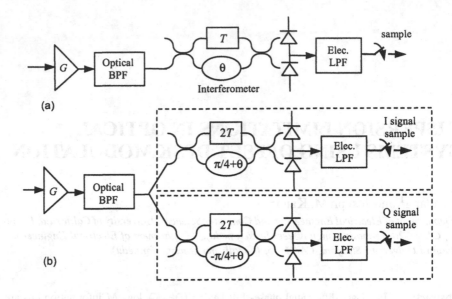

Figure 1.   Schematic diagram of optical receivers for (a) θ-2-DPSK and (b) θ-4-DPSK.

differences using Mach-Zehnder modulators (MZM). Recent transmission experiments using $\pi/2$-2-DPSK [5] showed that $\pi/2$-2-DPSK outperforms 0-2-DPSK in the presence of strong optical filtering and polarization-mode dispersion (PMD). In this paper, using quasi-exact numerical analysis, we investigate the performance of θ-2-DPSK and θ-4-DPSK in the presence of optical filtering, chromatic dispersion (CD) and first-order PMD when θ assumes arbitrary values. We show that for M-DPSK, $M = 2$ or 4, as θ increases from 0 to π/M, the tolerance to CD decreases; however, if strong optical filtering is used, the tolerance to first-order PMD can increase.

2.   SYSTEM MODEL AND BER CALCULATION

We assume that appropriate modulators are used to modulate θ-2-DPSK or θ-4-DPSK signals. The modulated signal is then passed through an optical bandpass filter, which corresponds to the multiplexer in a multi-channel system, and is launched into the optical fiber. The fiber is assumed to be lossless. In the fiber, the optical signal is affected by CD and first-order PMD. The impact of fiber nonlinearity is neglected in this work.

After transmission through the fiber, the signal is received by an appropriate receiver. Fig. 1 shows the designs of receivers for θ-2-DPSK and θ-4-DPSK. The received signal passes into a lumped optical amplifier, which adds amplified spontaneous emission (ASE) noise. We assume that ASE noise dominates over shot noise and thermal noise in the receiver. The amplifier output is filtered by an optical bandpass filter, and is passed into an optical delay-line Mach-Zehnder interferometer, which demodulates the received DPSK signal. We let T denote the information bit duration. In 2-DPSK receivers (Fig. 1(a)), the upper branch of the interferometer delays the optical signal by the symbol duration T and the lower branch shifts the phase of the optical signal by θ. In 4-DPSK receivers (Fig. 1(b)), a pair of interferometers is used to demodulate in-phase (I) and quadrature (Q) components of the signal, respectively. In the I interferometer, the upper branch delays the optical signal by the symbol du-ration $2T$, and the lower branch has an excess phase shift of $\pi/4 + \theta$. In the Q interferometer, the upper branch also delays the optical signal by $2T$, while the lower branch has an excess phase shift of $-\pi/4 + \theta$.

The bit-error ratio (BER) for DPSK in the presence of CD and PMD is calculated using the method proposed in [7, 8]. Briefly, this method expands the optical noise in a Fourier series using a Karhunen-Loeve series expansion (KLSE) and invokes a matrix formulation to express the decision sample in a non-central quadratic form of Gaussian random variables, which has a non-central chi-square distribution. The moment generating function of decision samples is obtained and the BER can be evaluated from it using the inverse Laplace transform and saddle point integration approximation. We refer to this procedure as the KLSE method, following [8] which describes in detail its application to DPSK systems with CD and first-order PMD. Although only the case $\theta = 0$ is considered in [8], the KLSE method can be easily adapted to arbitrary θ.

To verify the accuracy of the KLSE method for θ-2-DPSK and θ-4-DPSK, we compare the BERs obtained using the KLSE method to those obtained by Monte-Carlo simulation. The system configuration is as follows. The optical bandpass filters in both the transmitter and receiver are second-order Gaussian filter having equal 3-dB bandwidths given by B_o (full-width at half-maximum). The electrical lowpass filter is a fifth-order Bessel filter with a 3-dB cutoff frequency of B_e. The values of B_o, B_e and other system parameters are shown in Table 1.

Table 1.   Design parameters of θ-2-DPSK and θ-4-DPSK systems.

Parameters	θ-2-DPSK	θ-4-DPSK
Bit Rate R (Gb/s)	40	40
B_o (GHz)	50	25
B_e (GHz)	32	16
Duty ratio of elementary pulse	33%	33%
Noise figure of optical amplifier (dB)	3	3

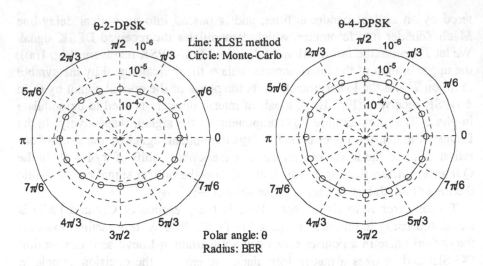

Figure 2.   Comparison of the BERs obtained by the KLSE method and by Monte-Carlo
simulation.

We assume the optical signal is launched into the fiber with equal projections into the two principal states of polarization, causing the worst-case PMD effect. First-order PMD is parameterized in terms of τ/T, the ratio of differential group delay (DGD) to bit duration. CD is characterized by the product R^2DL in the units of $(Gb/s)^2$ps/nm, where R is the bit rate, i.e., $R = 1/T$, D is the fiber CD parameter and L is the fiber length.

The BERs obtained using the KLSE method and Monte-Carlo simulation are presented in Fig. 2. In θ-2-DPSK systems, the transmitted signal power (at the modulator output) is 25 photons/bit, $\tau/T = 0.6$ and $D = 0$ ps/km/nm. In θ-4-DPSK systems, the transmitted signal power is also 25 photons/bit, $\tau/T = 1.2$ and $D = 0$ ps/km/nm. Fig. 2 shows that BERs computed by the KLSE method are in excellent agreement with Monte-Carlo simulation. Fig. 2 also shows that in θ-2-DPSK systems, the BER vs. θ curve is symmetrical with respect to the horizontal and vertical axes. In other words, θ-2-DPSK, $(-\theta)$-2-DPSK, $(\pi+\theta)$-2-DPSK and $(\pi-\theta)$-2-DPSK systems have identical BERs. For this reason, when we investigate the performance of θ-2-DPSK in next section, we consider $\theta \in [0, \pi/2]$ only. Furthermore, as the BER changes smoothly and monotonically as θ increases from 0 to $\pi/2$, we can focus on the three typical cases: $\theta = 0$, $\pi/4$ and $\pi/2$. In θ-4-DPSK systems, the BER vs. θ curve is nearly circular. In fact, it is symmetrical with respect to four axes: horizontal, vertical, $\theta = \pi/4$ and $\theta = 3\pi/4$. For this reason, we consider two typical cases for 4-DPSK: $\theta = 0$, $\pi/4$.

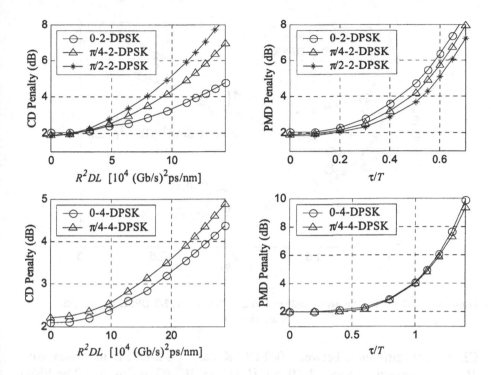

Figure 3. CD and PMD power penalty of θ-2-DPSK and θ-4-DPSK systems.

3. CD AND PMD PENALTIES FOR θ-2-DPSK AND θ-4-DPSK

In this section, we present the power penalties of θ-2-DPSK and θ-4-DPSK caused by CD and first-order PMD. The bit rate, elementary pulse shape, filter types, filter bandwidths and the noise figure of the optical amplifier are the same as in Section 2. The optical signal is still launched into the fiber with equal projections into the two principal states of polarization. We separately vary one of R^2DL and τ/T, while setting the other to zero, and calculate the power penalty at a BER of 10^{-9}. Note that our system design uses strong optical filtering, as in [5].

Fig. 3 presents the calculated power penalties at 10^{-9} BER for θ-2-DPSK and θ-4-DPSK. All the power penalties are referred to the quantum limits for 10^{-9} BER, which are 21.8 photons/bit and about 31 photons/bit for 2- and 4-DPSK, respectively. The decision threshold is always optimized to minimize BER; we find that the optimal threshold often deviates from zero by as much as 0.5% of the peak-peak excursion of the decision samples.

In the upper row of Fig. 3, we observe that for θ-2-DPSK, as θ increases from 0 to $\pi/2$, the CD penalty increases and the PMD penalty decreases. The

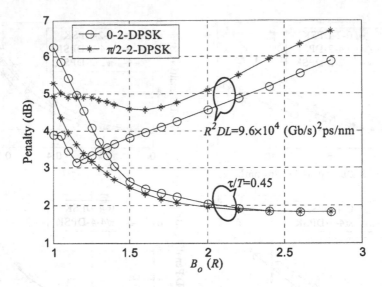

Figure 4.   CD and PMD power penalties of 0-2-DPSK and $\pi/2$-2-DPSK versus optical
bandwidth B_O.

CD penalty difference between 0-2-DPSK and $\pi/2$-2-DPSK increases with R^2DL, approaching about 2 dB for $R^2DL = 10^5$ (Gb/s)^{2}ps/nm. The PMD penalty difference between 0-2-DPSK and $\pi/2$-2-DPSK is about 1 dB for $\tau/T > 0.3$. These observations suggest that when strong optical filtering is used with 2-DPSK, the optimal choice of θ depends on the relative strengths of CD and PMD in a given system. An explanation of why 0-2-DPSK and $\pi/2$-2-DPSK have different tolerances to CD and PMD has been given in [9].

We now focus on the bottom row of Fig. 3. We observe that for θ-4-DPSK, as θ increases, the CD penalty increases and the PMD penalty decreases, similar to the case of θ-2-DPSK. However, the variation of CD and PMD penalties is not significant: 0-4-DPSK has only 0.4 dB lower CD penalty than $\pi/4$-4-DPSK. These two techniques have almost identical PMD penalties when $\tau/T < 1$. When $\tau/T > 1.2$, $\pi/4$-4-DPSK has a PMD penalty about 0.4 dB lower than 0-4-DPSK.

When the bandwidth of the optical filters is increased, the CD and PMD penalties of θ-M-DPSK system is changed. We consider θ-2-DPSK as an example. Fig. 4 shows the CD penalties at $R^2DL = 9.6 \times 10^4$ (Gb/s)^{2}ps/nm and the PMD penalties at $\tau/T = 0.45$ versus the bandwidth B_o of the optical filters. Focusing on the penalty difference between 0-2-DPSK and $\pi/2$-2-DPSK instead of the individual penalties, we see that the difference in PMD penalties decreases as B_o increases, and becomes negligible when $B_o$exceeds $2.25R$. By contrast, there is a substantial CD penalty difference between 0-2-DPSK and $\pi/2$-2-DPSK, although the penalty decreases for large B_o. Therefore, with weak optical filtering, 0-2-DPSK is preferable over $\pi/2$-2-DPSK, because the two techniques have similar PMD penalties, while the former technique has a smaller CD penalty.

4. CONCLUSIONS

We evaluated the CD and PMD power penalties of θ-2-DPSK and θ-4-DPSK systems. For θ-M-DPSK ($M = 2$ or 4), with either weak or strong optical filtering, when θ increases from 0 to π/M, CD power penalty increases. In the presence of strong optical filtering, when θ increases from 0 to π/M, PMD power penalty will decrease. However, the variation of CD and PMD penalties with θ is not significant in θ-4-DPSK system.

ACKNOWLEDGMENT

This research was supported, in part, by National Science Foundation Grant ECS-0335013.

REFERENCES

[1] J. H. Winters, "Differential detection with intersymbol interference and frequency uncertainty," *IEEE Trans. Commun.*, vol. COM-32, pp. 25–33, Jan. 1984.

[2] Korn, "Offset DPSK with differential phase detector in satellite mobile channel with narrow-band receiver filter," *IEEE. Trans. on Vehicular Technol.*, vol. 38, pp. 193-203, Nov. 1989.

[3] A. H. Gnauck, G. Raybon, S. Chandrasekhar, J. Leuthold, C. Doerr, L. Stulz, A. Agarwal, S. Banerjee, D. Grosz, S. Hunsche, A. Kung, A. Marhelyuk, D. Maywar, M. Movagassaghi, X. Liu, C. Xu, X. Wei, D. M. Gill, "2.5Tb/s (64×42.7 Gb/s) transmission over 40×100 km NZDSF using RZ-DPSK format and all-Raman-amplified spans," in *Tech. Digest of Postdeadline Papers, OFC 2002*, pp. FC2.1-FC2.3.

[4] W. Christoph, L. Jochen, R. Werner, "RZ-DQPSK format with high spectral efficiency and high robustness towards fiber nonlinearities," in *Proc. ECOC, 2002*, paper 9.6.6.

[5] X. Wei, A. H. Gnauck, D. M. Gill, X. Liu, U.-V. Koc, S. Chandrasekhar, G. Raybon, J. Leuthold, "Optical $\pi/2$-DPSK and its tolerance to filtering and polarization-mode-dispersion," *IEEE Photon. Technol. Lett.*, vol. 15, no. 11, pp. 1639-1641, Nov. 2003.

[6] H. Kolgelnik, R. M. Jopson and L. E. Nelson, "Polarization-mode dispersion", in *Optical Fiber Telecommunication IVB systems and impairments* (Ivan P. Kaminow and Tingye Li, eds.), pp. 725-861, San Diego, CA: Academic Press, 2002.

[7] E. Forestieri, "Evaluating the error probability in lightwave systems with chromatic dispersion, arbitrary pulse shape and pre-and postdetection filtering," *J. Lightwave Technol.*, vol. 18, pp. 1493-1503, Nov. 2000.

[8] J. Wang, J. M. Kahn, "Impact of chromatic and polarization-mode dispersion on DPSK systems using interferometric demodulation and direct detection," *J. Lightwave Technol*, vol. 22, no. 2, pp. 362-371, Feb. 2004.

[9] J. Wang, J. M. Kahn, "Conventional DPSK vs. symmetrical DPSK: comparison of dispersion Tolerances", *IEEE Photon. Technol. Lett.*, vol. 16, pp. 1397-1399, May 2004.

INTEGRATED OPTICAL FIR-FILTERS FOR ADAPTIVE EQUALIZATION OF FIBER CHANNEL IMPAIRMENTS AT 40 GBIT/S

Marc Bohn[1,2], Werner Rosenkranz[2], Folkert Horst[3], Bert Jan Offrein[3], Gian-Luca Bona[3], Peter Krummrich[4]

[1]*Siemens AG, Corporate Technology, Munich, Germany, marc.bohn@siemens.com;* [2]*University of Kiel, Chair for Communications, Germany, wr@tf.uni-kiel.de;* [3]*IBM Research GmbH, Zurich Laboratory, Switzerland, fho@zurich.ibm.com;* [4]*Siemens AG, ICN, Munich, Germany*

Abstract: In high bitrate optical transmission systems the dynamic changes of the transmission channel easily exceed the system tolerances for an error free operation. To meet the tolerances an adaptive equalizer is necessary. We demonstrate the capabilities of planar lightwave circuit integrated optical FIR-filters for an adaptive compensation of optical fiber channel impairments with electrical spectrum monitoring as feedback in simulations and measurements at 40 Gb/s.

Key words: optical communication; optical filters; optical equalizer; FIR filters; planar lightwave circuit; adaptive equalization; dispersion compensation; PMD compensation.

1. INTRODUCTION

The adaptive distortion compensation of fiber channel impairments in current and next generation high speed optical transmission systems is of high interest. With increasing bitrates and increasing complexity of the optical layer, signal distortions are increasing, while the tolerances of the system are decreasing. The dynamic changes of the transmission channel due to chromatic dispersion (CD), polarization mode dispersion (PMD) and nonlinear distortions, such as self-phase modulation (SPM), easily exceed the tolerable amount for an error free operation of the transmission system. To compensate for these time and frequency varying distortions and to meet the system tolerances, a static compensation approach is not sufficient anymore and an adaptive solution for equalization is necessary.

Adaptive equalization schemes exist in the electrical and in the optical domain. The electrical equalizers operate behind the opto-electrical conversion. Due to the envelope demodulation of the photo diode, the carrier and phase information get lost. Implemented filter structures are finite impulse response (FIR) and decision feedback (DF) [1] equalizers operating up to a bitrate of *10 Gb/s*. A new approach is maximum likelihood sequence estimation (MLSE) [2], but not yet implemented for high bitrates.

Optical equalizers have advantages in comparison to electrical ones, as they are not bitrate limited due to the electronics and operate in front of the nonlinear photodiode. Reported adaptive optical equalization experiments compensate either for chromatic dispersion or polarization mode dispersion by devices that model the inverse system, e.g. CD compensation by fiber bragg gratings (FBG) and etalons or PMD compensation by cascaded polarization control and birefringent elements [3-6].

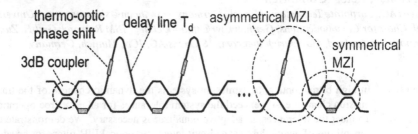

Figure 1.   Schematic structure of an integrated optical FIR-lattice filter: cascaded symmetrical and asymmetrical Mach-Zehnder Interferometers (MZI).

Within our approach, a planar lightwave circuit (PLC) integrated optical FIR-filter offers a large variability in compensating for distorting effects. These filter structures have variable complex coefficients and the transfer function can be arbitrarily tuned. Not only a single fiber impairment e.g. CD, SPM, or PMD can be compensated for, but also combinations of all these distortions [7-9]. In addition, optical FIR-filters have a periodic frequency response and tunable center wavelength. By matching the frequency response periodicity (i.e. the free spectral range (FSR)) to the channel grid, a single filter can equalize a number of WDM channels simultaneously. An efficient way of implementation is a lattice filter, which consists of cascaded symmetrical and asymmetrical Mach-Zehnder Interferometers (MZI), Fig. 1.

The device we used for the experiments is a 6th order lattice filter with a *FSR=100GHz*. It it designed and fabricated using the IBM high-index-contrast SiON technology [10]. The die size is *16×12 mm*.

Apart from the equalizer itself, a feedback signal for the automatic adaptation is necessary. Criteria for the adaptive control have been proposed in the time as well as in the frequency domain, e.g. eye opening, Q-factor, bit error rate or intersymbol interference and vestigial side band filtering, narrow optical filtering, monitoring a subcarrier, the clock intensity or the electrical spectrum [7–9,11–15].

In the time domain the classical adaptive equalization scheme in communication theory, the minimization of the intersymbol interference (ISI) with a Least Mean Square (LMS) error algorithm is a promising approach [7]. But the ISI generation is problematic at high bitrates and not yet implemented.

To demonstrate the adaptive capabilities of the optical FIR-filter, we choose electrical spectrum monitoring as adaptive feedback for simultaneous CD and PMD compensation, because of its sensitivity to both, good correlation to the signal distortions, simplicity and ease of implementation [8,9].

2. ADAPTIVE COMPENSATION OF CD AND PMD WITH ELECTRICAL SPECTRUM MONITORING AS FEEDBACK SIGNAL

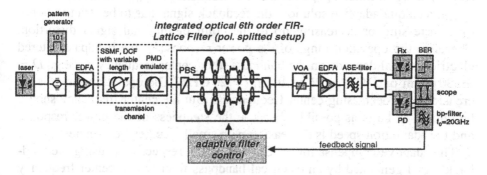

Figure 2.   System setup: adaptive equalization with optical FIR-filters and electrical spectrum monitoring. When compensating for PMD, this polarization splitted setup is necessary. Each polarization is equalized with a separate filter. For CD compensation, a single filter without polarization splitting is sufficient.

In contrast to the time and optical frequency domain criteria, adaptive feedback solutions within the electrical spectrum are fast, inexpensive and easy to implement by electrical bandpass filters and power monitoring. We demonstrate a strategy of monitoring a single frequency for combined adaptive CD and PMD compensation in a *40 Gb/s NRZ* transmission setup, Fig. 2.

Signal distortions are determined through changes of the transfer function of the optical fiber. The optical fiber transfer function is reflected in the power spectral density or the electrical spectrum of the received signal. Monitoring the power of the bandpass filtered received electrical signal for various dispersion values results in an oscillating characteristic with a global maximum ($f_0 < f_{bit}$) or minimum ($f_0 = f_{bit}$) at zero chromatic dispersion for linear

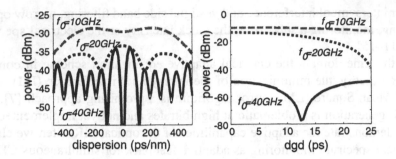

transmission, Fig. 3a. The bandpass filtered electrical PMD signal has well defined alternating power maxima and minima at a differential group delay (DGD) of e.g. n-times $T_{bit}/2$ for a center frequency of $f_0 = 40\,$GHz, Fig. 3b.

For a unique adaptive solution, the feedback signal has to be strictly monotone increasing or decreasing up to the point of minimal signal distortion. Therefore the operation range of the power response of the bandpass filtered electrical signal depends on the bandpass filter center frequency f_0, e.g. $D = \pm 180\,$ps/nm or $DGD = 25$ ps for $f_0 = 20\,$GHz. The operation range is increasing with decreasing center frequencies. But the center frequency should be chosen as large as possible, because the steepness of the power response and the adaptation speed is decreasing with smaller center frequencies.

The adaptive compensation of CD and PMD, respectively, using the feedback signal generated by an electrical bandpass filter with a center frequency $f_0 = 20\,$GHz is demonstrated in two experiments, Fig. 4a and b.

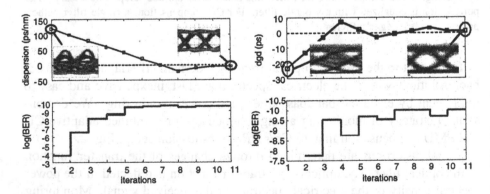

Figure 4.   Adaptive compensation results: (left) residual CD of the transmission channel & equalizer, eye pattern, BER; (right) residual DGD of the transmission channel & equalizer, eye pattern, BER.

From a starting point of approximately $D = 120\,\text{ps/nm}$ residual dispersion ($BER < 10^{-4}$) at the receiver, the adaptive control algorithm varies the residual dispersion of the equalizer until the optimum CD value is reached. For CD compensation only, a single filter is sufficient. The polarizations do not have to be splitted as indicated in Fig. 2. In a few iteration steps the eye pattern diagram is well opened, the bit error rate reduced and the dispersion compensated.

For PMD compensation, the orthogonal polarization modes have to be splitted, each polarization equalized by a separate filter and the polarizations combined (polarization splitted setup), see Fig. 2. The starting point is a DGD value of 23 ps at the receiver. The adaptive equalizer compensates for the DGD in a few iteration steps, the eye pattern diagram is well opened, the bit error rate reduced.

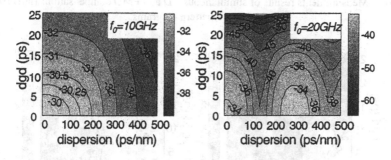

Figure 5. Power response of the received bandpass filtered electrical signal for distortions due to CD and PMD at 40 Gb/s: bandpass center frequency $f_0 = 10\,\text{GHz}$ (left), $f_0 = 20\,\text{GHz}$ (right).

For the combination of both signal distortions, PMD and CD, the bandpass filtered electrical signal power is a two dimensional function depending on the amount of CD and DGD with a global maximum at zero CD and DGD, Fig. 5. Therefore, electrical spectrum monitoring can be used to compensate for CD and DGD simultaneously. The simultaneous equalization of CD and PMD is shown experimentally, Fig. 6, and by simulations, Fig. 7.

In the measurement the transmission channel is set in a first step to a GVD value of $100\,\text{ps/nm}$. While compensating for GVD only, a sensitivity gain of 4.5 dB and a sensitivity penalty of less than 1 dB in comparison to the back to back case at a $BER = 10^{-9}$ is measured. Next, a PMD setting of the transmission channel of $DGD = 25\,\text{ps}$ is compensated. The initially closed eye pattern is clearly opened and the resulting sensitivity penalty is approximately 1 dB. Finally the transmission channel is set to GVD and PMD values of $GVD = 100\,\text{ps/nm}$ and $DGD = 25\,\text{ps}$. Equalizing the combination of PMD and GVD, the completely closed eye pattern is clearly opened and the sensitivity penalty is less than 1.5 dB.

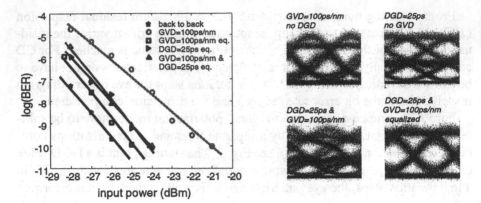

Figure 6.   Measurements results of simultaneous CD and PMD compensation: (left) BER, (right) eye pattern diagrams.

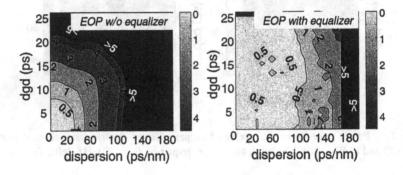

Figure 7.   Simulation results of simultaneous CD and PMD compensation: (left) w/o equalizer, (right) with adaptive equalizer.

The limits of this setup can be demonstrated by looking at the simulated eye opening penalties (EOP). The characteristic 1 dB EOP line is increased from $D = 60\,\text{ps/nm}$ and $DGD = 12\,\text{ps}$ without equalizer up to $D = 140\,\text{ps/nm}$ and $DGD = 24\,\text{ps}$ with equalizer. Exceeding the operation range leads the adaptive algorithm to converge into a local maximum. To increase the operation range, the bandpass filter frequency has to be decreased or combined with a bandpass filters of lower center frequency. For better equalization results inside the adaptation region, the filter order has to be increased.

3.   CONCLUSION

PLC integrated optical FIR-filters structures are a promising optical device to adaptively compensate for a single or combinations of several fiber channel impairments. The footprint is very small, and the filter coefficients, i.e the transfer function, can be fast and easily tuned by the thermo-optic effect. Electrical spectrum monitoring is a powerful, robust, fast and easy to implement solution for the adaptive feedback with a good correlation to signal distortions due to CD and PMD. Additional distortions due to SPM or chirp will be compensated by balancing the impact of SPM and residual CD into an optimum.

REFERENCES

[1] S. Otte et al., "Performance of electronic compensators for chromatic dispersion and SPM," *ECOC 2002*, Munich, paper 3.2.6.

[2] H. Bülow et al., "Electronic PMD mitigation – from linear equalization to maximum-likelihood detection," *OFC 2001*, Anaheim, paper WAA3-2.

[3] M. Sumetsky et al., "High-performance 40Gb/s fibre bragg grating tunable dispersion compensator fabricated using group delay ripple correction technique," *Electron. Lett.*, vol. 39, no. 16, pp. 1196–1198, 2003.

[4] D. J. Moss et al., "Multichannel tunable dispersion compensation using all-pass multi-cavity etalons," *OFC 2002*, FA6.

[5] F.Horst et al., "Tunable ring resonator dispersion compensators realized in high refractive index contrast SiON technology," *ECOC 2000*, paper PD2.2.

[6] R. Noe et al., "Polarization mode dispersion compensation at 10, 20, 40 Gb/s with various optical equalizers," *J. Lightwave Technol.*, vol. 17, pp. 1602–1616, 1999.

[7] M. Bohn et al., "An adaptive optical equalizer concept for single channel distortion compensation," *ECOC 2001*, paper Mo.F.2.3.

[8] M. Bohn et al., "Simultaneous adaptive equalization of group velocity and polarization mode dispersion at 40 Gb/s with integrated optical FIR-filters and electrical spectrum monitoring as feedback," *ECOC 2003*, paper Th.2.2.4.

[9] M. Bohn et al., "Experimental verification of combined adaptive PMD and GVD compensation in a 40 Gb/s transmission using integrated optical FIR-filters and spectrum monitoring," *OFC 2004*, Los Angeles, paper TuG3.

[10] G.-L.Bona et al., "SiON high refractive-index waveguide and planar lightwave circuits," *IBM J. Res.&Dev.*, vol. 47, no. 2/3, pp.239–249, 2003.

[11] F.Buchali et al., "Fast eye opening monitor for 10Gb/s and its application for optical PMD compensation," *OFC 2001*, Anaheim, paper TuP5.

[12] Q. Yu et al., "Chromatic dispersion monitoring technique using sideband optical filtering and clock phase-shift detection," *J. Lightwave Technol.*, vol. 20, no. 12, pp. 2267–2271, 2002.

[13] C. K. Madsen, "Chromatic and polarization mode dispersion measurement technique using phase sensitive detection," *OFC 2001*, Anaheim, paper Mo6-1.

[14] M.N.Petersen et al., "Dispersion monitoring and compensation using a single inband subcarrier tone," *OFC 2000*, Baltimore, paper WH4-1.

[15] A.Sano et al., "Extracted-clock power level monitoring scheme for automatic dispersion equalization in high speed optical transmission systems," *IEICE Trans. Commun.*, vol. E84B, no. 11, pp. 2907–2914, 2001.

PERFORMANCE OF ELECTRONIC EQUALIZATION APPLIED TO INNOVATIVE TRANSMISSION TECHNIQUES

Vittorio Curri, Roberto Gaudino, and Antonio Napoli
OptCom Group, PhotonLab
Dipartimento di Elettronica, Politecnico di Torino
C.so Duca degli Abruzzi 24, 10129, Torino, Italy
www.optcom.polito.it - optcom@polito.it
Tel. +39.011.2276301 - Fax +39.011.2276299

Abstract: We investigate in this article the use of electronic equalization on dispersion limited systems for different modulation formats. Besides analyzing equalization on standard NRZ intensity modulation as a reference, we focus on two advanced modulation formats, Duobinary and Differential Phase Shift Keying (DPSK) which are recently gaining large attention. We demonstrate that the introduction of an electronic equalizer strongly improves standard NRZ performance, whereas it has a limited effect on Duobinary and DPSK formats. Moreover, we give rules for the optimal choice of the equalizer transversal filter parameters, i.e., the number of taps and the delay between taps.

Key words: Q-factor; electronic equalization; FIR filters; modulation formats; intersymbol interference; chromatic dispersion; optical communication.

1. INTRODUCTION

The use of electronic equalization as a way to mitigate the effects of Inter-Symbol Interference (ISI) in traditional communication systems (i.e., wireless and copper wireline systems) has been deeply studied, and a large literature is available [1, 2]. In general, all these techniques are based on the use of a transversal digital filter (which acts as a Finite Impulse Response (FIR) filter) at the receiver with a proper choice of taps parameter.

In early nineties, Winters *et al.* [3] proposed for the first time the use of electronic equalization in optical communication systems. This approach has recently gain momentum, as it is today seen as a cheaper (even though less performing) alternative to all-optical equalization. Several solutions have been proposed, mainly in order to reduce the system impact of chromatic dispersion (CD) [4, 6] and Polarization Mode Dispersion (PMD) [5].

Electronic compensation in optical systems still lacks a complete theoretical understanding. This is mainly due to the fact that the analysis carried out for equalization of linear channels in traditional communication systems [1, 2] cannot be directly applied to the optical environment, where CD and PMD are linear at the "optical" level, but (particularly for CD) become nonlinear at the receiver side after the photodiode "square-law" conversion [4–6].

In this article, we focus on CD electrical compensation for system limited by Amplified Spontaneous Emission (ASE) noise, focusing on advanced modulation formats. Recently, alternative modulation formats have been proposed in order to overcome linear and non-linear propagation limits of standard NRZ. Among these formats, the most promising for an advantageous implementation seem to be the Differential Phase Shift Keying (DPSK) [7] and the optical Duobinary (DB) [8]. Anyway, the combination of advanced modulation formats with electronic equalization has not been deeply studied yet, at least to our best knowledge.

The purpose of this work is thus twofold. On one side, we analyze the impact of electronic equalization on DB and DPSK modulation formats, comparing the results to standard NRZ, which is taken as a reference. On the other side, we present a set of results on the optimization of number of taps of the equalizer transversal filter and delay between taps (for each modulation format).

The paper is organized as follows. In Section 1.1 we describe the reference scenario and the equalizer scheme together with the algorithm used for the evaluation of the FIR filter coefficients. In Section 2 we present the simulation results, demonstrating that the electronic equalization may strongly improve NRZ performance, but has a limited effect on systems based on DB and DPSK modulation formats. Finally, in Section 3, we draw the conclusions and envision possible evolutions of this work.

1.1 System setup

We analyzed and simulated [9] the system scenario shown in Fig. 1, corresponding to an externally modulated system limited by the accumulation of ASE noise and by CD-induced intersymbol interference, while we neglect other effects, such as fiber nonlinearities. These assumptions are reasonable for high capacity metro and extended metro links using SMF fibers, a scenario that is today gaining increasing attention, due to the new optical market trends.

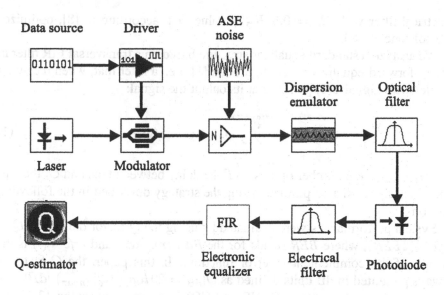

Figure 1. Block diagram of the analyzed system setup.

We considered a $R_b = 10$ Gbit/s systems using a chirp-free external Mach-Zehnder amplitude Modulator (MZM) at the transmitter side. For the three considered modulation formats, the MZM operates according to different driving schemes. Let V_{off} (V_{on}) be the driving voltage for a minimum (maximum) of the MZM transfer function, and $V_\pi = V_{on} - V_{off}$ the resulting on-off voltage swing. In order to produce standard NRZ modulation, the MZM is simply driven between V_{off} and V_{on} voltages. For the DPSK modulation format, the input digital stream is first differentially precoded, then the MZM is driven between the two voltages $V_{off} \pm V_\pi$ [2]. This scheme yields a two level phase modulation associated to "0" and "π" radians. To obtain a DB modulation, the input digital stream is processed by the duobinary precoder and then filtered using a narrow lowpass electric filter (5 pole Bessel filter with -3 dB bandwidth equal to $0.25 \cdot R_B$) [8]. The filtered sequence is used to drive the MZM between the two voltages $V_{off} \pm V_\pi$ [8].

The modulated optical signal is sent to an optical channel affected by CD and ASE noise. For all modulation formats, results will be presented as a function of the Optical Signal to Noise Ratio (OSNR) (calculated over an optical bandwidth of 50 GHz) at the receiver input.

We assumed to use a standard optical receiver, based on a 2nd order Super-Gaussian optical filter, having a -3 dB bandwidth equal to $B_o = 40\ GHz$, an ideal photodiode and a typical Bessel 5th order electrical filter -3 dB bandwidth $B_e = 0.75\ R_B$ for NRZ and DPSK. For the DB format only, we used an electrical filter with $B_e = 0.5 \, R_B$, a value that, according to [8], optimizes DB tolerance to CD.

We analyze a standard equalizer structure based on a transversal FIR filter in a feed-forward equalization configuration [1], i.e., a filter that, when receiving an electrical signal $x(t)$ generate at its output the signal:

$$y(t) = \sum_{i=0}^{N_{taps}-1} C_i \cdot x(t - i\Delta T) \tag{1}$$

where N_{taps} is the number of taps, ΔT the delay between taps and C_i the taps coefficients, which are optimized using the strategy described in the following section.

System performances were evaluated by using the Q-factor defined as $Q = erfc^{-1}(2 \, BER)$, where BER stands for the *Bit Error Rate* and $erfc^{-1}(\cdot)$ is the inverse of the complementary error function. In this paper, the Q factor is always presented in dB units defined as $Q|_{dB} = 20 \, log_{10} \left(Q|_{linear} \right) \, [dB]$. We spanned system parameters (OSNR and CD) in order to cover the 12-18 dB range for the Q-factor.

2.   SIMULATION AND RESULTS

As an optimization strategy, we used standard numerical optimization routines (such as the *Quasi-Newton method* [10] and the *simplex search method* [11] implemented inside the commercial software Matlab^{TM} based on finding, for each system realization, the vector of taps coefficients C_i that maximizes the Q-factor, i.e., that minimizes the BER.

Besides the tap coefficients, we also investigate different values for the delay ΔT and the number of coefficients N_{taps}. In particular, we explored $\Delta T = T_B, T_B/2$ and $T_B/3$ (where $T_B = 1/R_B$ is the bit duration) and $N_{taps} = 3, 7,$ 9 and 15.

For each system setup, we performed an accurate time-domain simulation [9] of the signal propagation, then we optimized the equalizer FIR coefficient and obtain the resulting Q-factor. The simulated digital stream was a $2^{13} - 1$ PRBS sequence. We explored accumulated chromatic dispersion values D_{acc} in the [0, 4600] ps/nm range, and $OSNR \in [9, 15]$ dB range. We presented all results showing the trace connecting all points giving rise to a Q-factor equal to 17 dB (corresponding to $BER \sim 10^{-12}$) in the D_{acc} and $OSNR$ plane. Basically, we show a plot of the $OSNR$ required to get $Q = 17$ dB for each value of D_{acc}. We compared each system with and without equalizer.

We start by investigating, for the standard NRZ format, the influence of the tap delay ΔT and the number of taps N_{taps}. Results are shown in Fig. 2; the different curves refer to the unequalized system, and to the equalized system for $N_{taps} = 15$ (i.e., an extremely long transversal filter) and tap delays $\Delta T = $

$T_B, T_B/2$ *and* $T_B/3$. Contrary to the typical results shown for example in [1], we show that $\Delta T = T_B$ gives poor performances, while a great improvement is obtained by using $\Delta T = T_B/2$. Moving to $\Delta T = T_B/3$ gives only marginal improvement, showing that for our considered setup $\Delta T = T_B/2$ is a good choice. We explain this results as follows. Most standard studies on equalization [1, 2] assumes that a "matched" filter precedes the equalizer. In most cases an integrate & dump filter with integration time T_b is assumed as a reference. In this situation, choosing a delay $\Delta T < T_B$ is useless, since the transversal filter acts on samples that are already correlated on a T_b time window by the filter *preceding* the equalizer. On the contrary, in our scenario we assumed typical optical and electrical filter that are always significantly larger than a "matched" filter, and thus correlate the signal on a time window smaller than T_B, consequently justifying a tap delay smaller than T_b.

After setting $\Delta T = T_B/2$, we investigated the performance for different N_{taps} values. We show in Fig. 3 the results for the cases $N_{taps} = 3, 7, 11$ and 15. This figure shows that a FIR equalizer with more than 7 taps does not increase performances. We notice here that a filter with 7 taps and $\Delta T = T_B/2$ has a total "memory" equal to approx. 4 bits. We observed that for high dispersion values, the seven taps coefficients are approximately symmetric with respect to the central taps, as can be expected from the symmetric nature of the dispersion impulse response. This physically means that the equalizer "handles" a time window with a range $\pm 2T_b$ around the sampling time of each bit.

Using these optimized parameters $N_{taps} = 7$ and $\Delta T = T_B/2$, Fig. 3 shows that electronic equalization can greatly improves performance for the standard NRZ format. We define the CD limit as the value giving a 2-dB OSNR penalty with respect to the back-to-back configuration; in Fig. 3 it can be observed that the back-to-back configuration requires an OSNR=11 dB (in order to give a Q-factor equal to 17 dB), so that the 2-dB OSNR penalty corresponds to the points of the curve that crosses the OSNR=13 dB level. Using this reference, we observe an improvement from 750 ps/nm in the unequalized case to 1550 ps/nm in the equalized case, i.e, approximately a doubling of the dispersion limit.

Finally, we compared in Fig. 4 all the considered modulation formats with (dashed lines) and without (solid lines) equalization. This is the fundamental result of this article, showing that, for DPSK and DB, electronic equalization does not significantly increase performance. Using the same reference introduced for NRZ, it can be seen that electronic equalization improves the dispersion limit by only 250 ps/nm for both DPSK and DB. The limited benefit given by electronic equalization on these two modulation formats is not easy to be intuitively interpreted. A first observation is that both formats generates an electrical signal that, even *without* chromatic dispersion, shows a strong correlation over a one bit time window. This is due to the very narrow filtering

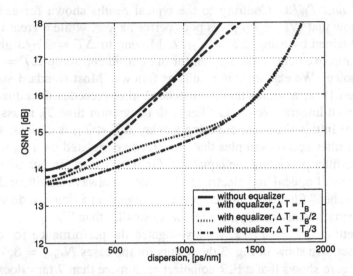

Figure 2. $OSNR$ vs. D_{acc} @ $Q = 17$ dB for the NRZ-OOK format. Different dotted curves refers to $\Delta T = T_B$, $T_B/2$, and $T_B/3$, and $N_{taps} = 15$. The unequalized curve is shown as comparison.

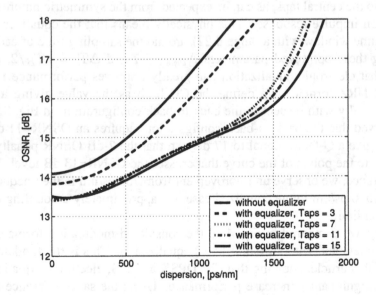

Figure 3. $OSNR$ vs. D_{acc} @ $Q = 17$ dB for the NRZ-OOK format. Different dotted curves refers to $N_{taps} = 3$, 7, 11 and 15, and $\Delta T = T_B/2$. The unequalized curve is shown as comparison.

at the transmitter side for duobinary, and by the receiver interferometric filter for DPSK. Moreover, in the case of duobinary, the format is *intrinsically* ex-

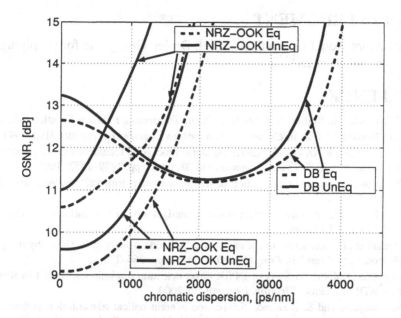

Figure 4. $OSNR$ vs. D_{acc} @ $Q = 17$ dB for the NRZ-OOK, DPSK and DB format. Equalized results are obtained with $N_{taps} = 7$, and $\Delta T = T_B/2$ for NRZ-OOK and DPSK, and $N_{taps} = 15$, and $\Delta T = T_B/3$. The unequalized curves are shown as comparison.

tremely robust to CD, as it has been shown in many papers, such as [8]. In fact, observing Fig. 4, it can be noted that the DB format, without equalization, reaches approximately the same performance of the back-to-back NRZ ($Q = 17$ dB with only $OSNR = 11.2$ dB) for $D_{acc} = 2200$ ps/nm. The corresponding eye diagram is this situation is nearly ISI-free, and thus it cannot be significantly improved by equalization.

Finally, we notice that for all formats, and even without dispersion i.e., in the back-to-back configuration, equalization improves performance of approx. 0.5 dB with respect to the unequalized system. This result can be explained by observing that, since our strategy optimizes taps coefficients using Q/BER as a target, it minimizes both ISI *and* noise. Thus, the resulting receiver equivalent filtering function tends to be closer to an ideally matched filter in the equalized case.

3. CONCLUSIONS

For the first time, to our best knowledge, we presented a comprehensive analysis of electronic equalization in connection to advanced modulation formats, showing that the use of electronic equalization based on FIR filter, while giving great benefit to standard NRZ, gives only limited advantages for DB and DPSK.

ACKNOWLEDGMENT

The authors would like to thank RSoft Design Group, Inc for supplying the simulation tool OptSim^{TM}.

REFERENCES

[1] S. Benedetto, E. Biglieri, V. Castellani, *Digital Transmission Theory*, Prentice Hall, 1987.

[2] J. G. Proakis, M. Salehi, *Communication Systems Engineering*, Prentice Hall, 1994.

[3] J. H. Winters, R. D. Gitlin, "Electrical signal processing techniques in long-haul fiber-optic systems", *IEEE Trans. Commun.*, vol. 38, no. 9., pp. 1439–1453, 1990.

[4] H. Bülow, "Electronic equalization of transmission impairments," in *Proc. OFC 2002*, pp. 24–25.

[5] H. Bülow et al., "Adaptation of an electronic mitigator by maximization of the eye opening," *ECOC'00*, Munich.

[6] Buchali et al., "Reduction of the chromatic dispersion penalty at 10 Gbit/s by integrated electronic equalisers," in *Proc. OFC 2000*, vol. 3, pp. 268–270.

[7] M.Rohde et al. "Robustness of DPSK direct detection transmission format in standard fiber WDM systems," *Electron. Lett.*, vol. 36, 2000.

[8] K. Yonegana and S. Kuwano, "Dispersion tolerant optical transmission system using duobinary transmitter and binary receiver," *J. Lightwave Technol.*, vol. 15, pp. 1530–1537, 1997.

[9] www.rsoftdesign.com

[10] Broyden, C.G., "The convergence of a class of double-rank minimization algorithms," *Journal Inst. Math. Applic.*, vol. 6, 1970.

[11] J.C. Lagarias et al., "Convergence properties of the Nelder-Mead simplex method in low dimensions," *SIAM Journal of Optimization*, vol. 9, 1998.

PERFORMANCE BOUNDS OF MLSE IN INTENSITY MODULATED FIBER OPTIC LINKS

N. Alić, G. C. Papen, L. B. Milstein, P. H. Siegel, and Y. Fainman
Department of Electrical and Computer Engineering, Jacobs School of Engineering, University of California San Diego, La Jolla, CA 92093-0407. (nalic@ece.ucsd.edu)

Abstract: Theoretical bounds for maximum likelihood sequence estimation (MLSE) for intensity modulated fiber-optic links using an integrate and dump receiver structure are derived. These bounds are used to compare performance of different modulation formats. We show that while significant performance differences exist when bit-by-bit detection is used, the performance differences between modulation formats are less significant when using a MLSE receiver structure.

Key words: maximum likelihood estimation; intersymbol interference; modulation formats; optical fiber communication.

1.   INTRODUCTION

The origin of most intersymbol interference (ISI) in single mode fiber optic links is the frequency dependence of the index of refraction that results in Group Velocity Dispersion (GVD) [1-2]. Currently, almost all deployed fiber optic links employ incoherent on-off keying (OOK) that relies on the detection of the presence of optical energy (Mark) or the absence (Space) in a single bit slot. Consequently, the energy leakage from Marks to Spaces caused by ISI leads to performance degradation and is the motivation for the development of mitigation techniques.

The importance of GVD mitigation has attracted significant interest from the scientific and engineering communities. Both all-optical and all-electrical solutions have been proposed [2]. The vast majority of these techniques, such as dispersion compensating fiber and active dispersion compensating modules, rely on the physical characteristics of GVD. Electrical equalization techniques, on the other hand, are potentially lower in cost and more flexible. This paper considers the application of maximum likelihood sequence estimation (MLSE)

[3, 4] to mitigate GVD using a simple square-law detector coupled with an integrate-and-dump receiver structure that is used in most current optical communication systems.

MLSE has been previously considered for optical communications in [5] where a theoretically optimal receiver incorporating a combination of a bank of matched filters and a Viterbi Algorithm was suggested for the mitigation of polarization mode dispersion (PMD). Recently, a mathematical treatment of signal dependent noise in conjunction with MLSE was presented in [6]. However, it is currently not practical to employ a large number of reconfigurable filters in existing optical links where a simple integrate-and-dump filter is almost universally used. The use of an integrating filter implies that the detection process will be sub-optimal. The goal of this paper is to quantify, for the first time, the reduction in performance resulting from utilization of this practical, but suboptimal, filter in conjunction with MLSE in fiber optic links.

2.   MLSE IN FIBER OPTIC LINKS

In contrast to the RF communication systems that perform linear detection of amplitude and phase, optical communication systems employ non-coherent square-law detection where the output is proportional to the magnitude squared of the optical field. A standard MLSE algorithm cannot be directly applied for this type of receiver structure, but must be modified to make it efficient. We start by assuming an Additive White Gaussian Noise (AWGN) model where the noise in the channel is dominated by additive electrical noise. This model is valid for short haul and metropolitan links that either contain no optical amplifiers, or where the noise is dominated by the noise component originating in the detector and the electronics in the post-detection circuitry [7]. The AWGN noise model allows relatively simple and direct evaluation of the performance of MLSE for fiber optic communication links.

As discussed in [4, 5], the optimal receiver for an AWGN channel with a power spectral density $N_0/2$ where ISI spans m bits[1] consists of a set of 2^{2m+1} matched electrical filters. The outputs of these matched filters are then used as an input for the Viterbi Algorithm. Adopting the labeling from the system block diagram in Fig. 1, the log-likelihood function for the detected signal vector $\underline{r}$ after the integrator may be written as

$$\phi_{R|S_{\underline{a}}}(\underline{r}|\underline{s_a}) = \log\left[f_{R|S_{\underline{a}}}(\underline{r}|\underline{s_a})\right] = \frac{1}{N_0 T} \left\|\underline{r} - \underline{s_a}\right\|_2^2 \tag{1}$$

---

[1]The span of ISI is defined as the number of bit intervals on *one* side of a mark in which the integrated power after detection is larger than 1% of the power contained in a mark after propagation.

where $\underline{s}_a$ is a set of noiseless response vectors corresponding to an input vector $\underline{a}$, T is the signaling interval and $\|x\|_2^2 = \sum_i x_i^2$ is the Euclidean norm. The log-likelihood of Eq. (1) defines the metric that is used by the Viterbi Algorithm. The complete process can be interpreted as a search for a sequence of noiseless responses that is closest, in the Euclidian sense, to the received signal vector.

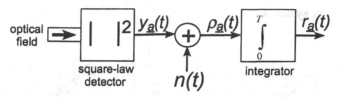

Figure 1. System block diagram

2.1 Error probability and performance

The use of the Viterbi Algorithm allows for a rigorous error analysis. With the Euclidean distance interpretation of Eq. (1), the event that the Viterbi Algorithm has chosen an incorrect sequence, i.e., other than the one actually transmitted, implies that the Euclidean distance from the received vector to an incorrect vector is smaller than that to the correct one. Thus, the probability that an incorrect sequence is preferred by the Viterbi Algorithm is related to the Euclidean distance between two vectors [4]

$$d^2(\underline{s}_1, \underline{s}_2) = \left\| \underline{s}_{a_1} - \underline{s}_{a_2} \right\|_2^2 .$$

Specifically, in the case of AWGN with a noise power $N_0 T/2$, we can express the pair-wise error probability between sequences $\underline{s}_{a_1}$ and $\underline{s}_{a_2}$ as

$$P\{\underline{s}_1 \to \underline{s}_2\} = P\{\phi_{\underline{a}_1 + \underline{e}} < \phi_{\underline{a}_1}\} = \frac{1}{2}\mathrm{erfc}\left(\frac{d(\underline{s}_1, \underline{s}_2)}{2\sqrt{N_0 T}}\right) \equiv \frac{1}{2}\mathrm{erfc}\left(\frac{d(\underline{e})}{2\sqrt{N_0 T}}\right) \tag{2}$$

where in the last expression, $\mathrm{erfc}(x)$ is the complementary error function. The binary error vector $\underline{e}$ satisfies the modulo 2 addition relationship $\underline{a}_2 = \underline{a}_1 \oplus \underline{e}$ that describes the error event of mistaking the vector $\underline{a}_1$ for $\underline{a}_2$. Bounds for the probability of error using the non-optimal filter in a high SNR regime may be derived following [4, 8-9] and can be written as

$$\frac{1}{2}\psi''(d_{\min})\mathrm{erfc}\left(\frac{d_{\min}}{2\sqrt{N_0 T}}\right) \leqslant P_e \leqslant \frac{1}{2}\psi'(d_{\min})\mathrm{erfc}\left(\frac{d_{\min}}{2\sqrt{N_0 T}}\right). \tag{3}$$

In Eq. (3), $d_{min} = \min\limits_{\underline{e} \in E} \{d(\underline{e})\}$ is the signal space minimum distance and $\psi'(d_{min})$ and $\psi''(d_{min})$ are coefficients that depend on the signal space configuration and d_{min}, but are independent of the noise characteristics. While the magnitudes of $\psi'(d_{min})$ and $\psi''(d_{min})$ affect the bounds, their contribution is only significant in a low SNR regime due to the sharp descent of the erfc function [4]. This low SNR regime is generally not used for fiber-based optical communication systems.

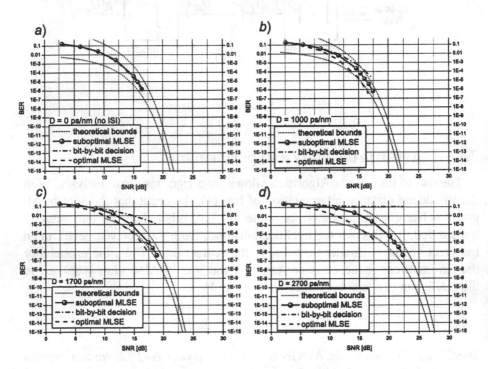

Figure 2. MLSE performance for different amounts of accumulated dispersion for NRZ format at 10 Gbps. Fiber propagation was simulated using the Virtual Photonics Transmission Maker software package. The solid dots are the results of simulation. The fine dashed lines are the bounds given in Eq.(3) for the suboptimal integrating receiver. The coarse dashed lines are simulations of an optimal MLSE receiver. The results of bit-by-bit detection are also shown.

Fig. 2 shows the performance of the NRZ format for four different amounts of accumulated dispersion, including the error bounds defined in Eq. (3). Signal space minimum distances were found by an exhaustive signal space search. As expected, MLSE outperforms the bit-by-bit integrate-and-dump receiver whenever ISI is present. It is noticeable that the performance power penalty increases with the amount of the accumulated dispersion and can become as large as 6 dB in Fig. 2d where a span of 5 bits of ISI results in a completely closed eye. The deterioration in performance closely matches the reduction in the value of d_{min}^2. As previously stated, the optimal detector would have to rely on a bank of 2^{2m+1} filters matched to each one of the possible channel responses for m bits of ISI. The output would subsequently be used by the Viterbi Algorithm. What is somewhat surprising is that there is a relatively small power penalty, (2-3 dB on average), for using the suboptimal integrating receiver structure (dots in Fig. 3) relative to the much more complicated matched filter receiver structure (coarse dashed line). The small power penalties are partially explained by noting that the information carrying waveforms with widths less than the symbol interval broaden as they propagate in the fiber. The impulse response of the fiber channel for these longer distances more closely matches that of the fixed integrating filter reducing the overall power penalty. This explanation is supported by the fact that the reduction in performance matches the reduction in the average SNR caused by the utilization of the mismatched filter.

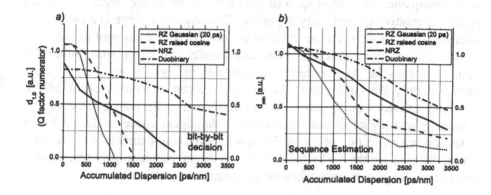

Figure 3. (a) Minimum distance d_{min} as function of the accumulated dispersion for bit-by-bit detection for four modulation formats. (b) Minimum distance d_{min} as function of the accumulated dispersion for sequence estimation for four modulation formats.

The preceding analysis can also be applied to the assessment of the performance of an MLSE applied to different modulation formats in comparison to the bit-by-bit decision. As already stated, for AWGN channels in a high SNR regime, d_{min} is the parameter that defines performance. On the other hand, the probability of error for bit-by-bit decision in AWGN is given by [10]

$$P_e = \frac{1}{2}\text{erfc}(Q/\sqrt{2}). \tag{4}$$

For the case of signal independent noise, $Q = (i_1 - i_0)/(2\sigma_0)$ where $\sigma_0 = \sqrt{N_0 T/2}$ is the RMS noise and $i_{1,0}$ are the average signals corresponding to a Mark and Space respectively. The performance is determined by the separation

$d_{1,0} = i_1 - i_0$ of the two symbols, and the expression for Q can be rewritten as $Q = (d_{1,0}/2)/\sigma_0$.

Fig. 3 shows the comparison in performance of MLSE performed for three popular modulation formats: RZ (two different pulse shapes), NRZ and the duobinary format [11]. Results were generated assuming equal average power for all four signal waveforms at the input to the fiber, and neglecting fiber attenuation to emphasize performance dependence on the waveform evolution. The received waveforms in the absence of noise were generated using VPI, while the signal space minimum distances were determined using well-established algorithms [12]. We note that, unlike the bit-to-bit case, where doubinary substantially outperforms other formats for values of the accumulated dispersion above 1000 ps/nm, when MLSE is performed, the difference in performance between the three formats is considerably reduced. In particular, the performance of the NRZ format approaches a doubinary format to within 2 dB. The cost of this modest power penalty is a more complex Viterbi Algorithm which is a consequence of a larger span of ISI for NRZ due to the larger transmitted spectrum relative to duobinary signaling. Consequently, there is a clear engineering trade-off between the transmitter complexity, the amount of launched power, and receiving structure. Note additionally that signal space minimum distance values for the two detection strategies in Figs. 3a and 3b start from approximately the same value. As a result, we can conclude that the MLSE approach provides the improvement in performance by taking advantage of operating in a higher dimensional space. The signal space of the bit-by-bit detection is a projection of the MLSE space onto a single dimension. The minimum distance for the bit-by-bit detection is therefore necessarily compressed and eventually goes to zero when the eye completely closes.

ACKNOWLEDGMENTS

The authors would like to acknowledge support from Applied Micro Circuits Corporation, UC Discovery Grant Program, and National Scientific Foundation through grant NSF-0123405.

REFERENCES

[1] G. P. Agrawal, *Nonlinear Fiber Optics*, chap. 3, San Diego: Academic Press, 2001.

[2] A. E. Willner and B. Hoanca, "Fixed and tunable dispersion compensation," in *Fiber Optic Telecommunication IV B* (I. Kaminow and T. Li, eds.), San Diego: Academic Pres, 2002.

[3] G. D. Forney Jr., "Maximum-likelihood sequence estimation of digital sequences in the presence of intersymbol interference," *IEEE Trans. Inform. Theory*, vol. IT-18, no. 3, May 1972.

[4] S. Benedetto and E. Biglieri, *Principles of Digital Transmission*, Kluwer Academic/Plenum Publishers, 1999.

[5] H. F. Haunstein et al., "Design of near optimum electrical equalizers for optical transmission in the presence of PMD," in *Proc. OFC 2001*, Paper WAA4-1, 2001.

[6] A. J. Weiss, "On the performance of electrical equalization in optical fiber transmission systems," *IEEE Photon. Technol. Lett.*, vol. 15, no. 9, pp. 1225–1227, 2003.

[7] B. L. Kasper et al., "High bit-rate receivers, transmitters and electronics," in *Optical Fiber Telecommunications IV A*, sect. 1.4.3.5, chap. 16, Elsevier Science, 2002.

[8] G. J. Foschini, "Performance bound for maximum-likelihood reception of digital data," *IEEE Trans. Inform. Theory*, vol. IT-21, no. 1, Jan. 1975.

[9] Y. J. Liu, I. Oka and E. Biglieri, "Error probability for digital transmission over nonlinear channels with application to TCM," *IEEE Trans. Inform. Theory*, vol. 36, no. 5, Sept. 1990.

[10] S. D. Personick, "Receiver design for digital fiber optic communication systems – I," *Bell Syst. Tech. J.*, vol. 52, no. 6, pp. 843–874, 1973.

[11] K. Yonenaga and S. Kuwano, "Dispersion-Tolerant Optical Transmission System Using Duobinary Transmitter and Binary Receiver," *J. Lightwave Technol.*, vol. 15, no. 8, pp. 1530–1537, Aug. 1997.

[12] M. G. Mullegan and S. G. Wilson, "An improved algorithm for evaluating trellis phase codes", *IEEE Trans. Inform. Theory*, vol. 30, pp. 846–851, 1984.

ON MLSE RECEPTION OF CHROMATIC DISPERSION TOLERANT MODULATION SCHEMES
Combining Chirped NRZ and Duobinary Transmission with EDC

Helmut Griesser[1], Joerg-Peter Elbers[1], and Christoph Glingener[2]

[1]*Marconi, Backnang, Germany;** [2]*Marconi, Coventry, United Kingdom*
{helmut.griesser | joerg-peter.elbers | christoph.glingener}@marconi.com

Abstract: Maximum likelihood sequence estimation (MLSE) in the electrical domain at the receiver of an optical fibre link is investigated as a means to improve the dispersion tolerance of the modulation formats chirped NRZ (CNRZ) and optical duobinary modulation (ODB).

Key words: optical fiber communication; electronic dispersion compensation; equalizers; maximum likelihood estimation; chirp modulation; optical duobinary modulation.

1.   INTRODUCTION

Already back in 1990 Winters [1] recognized the potential of electrical signal processing for reducing the distortions of optical fibre systems. As electrical equalisers are placed at the receiver after the optical front end, they can be easily combined with different modulation formats and optical measures that aim at an improved dispersion tolerance. However, whilst most of these methods have been investigated intensely, there is only very few information on the benefit of combining them with electrical signal processing. In [2] linear feed forward filter equalisation (FFE) of chirped transmission was considered, and

*The work reported in this paper has been supported in part by the German Ministry of Education and Research (BMBF) under contract number 01 BP 260. The authors are responsible for the content of this paper.

simulation results on a decision feedback equalizer (DFE) receiver for chirped and duobinary transmission are given in [3].

Maximum likelihood sequence estimation (MLSE) was shown in several publications to be superior over FFE and DFE type equalizers (see e.g. [4]). By now, with the advent of analogue/digital converters at 10 Gb/s and the progress of semiconductor development, the implementation of MLSE equalisers even at 10 Gb/s has come into reach. Therefore, in this paper the reception of chirped NRZ (CNRZ) and optical duobinary (ODB) modulation with an MLSE type receiver is investigated.

In the following section the basic conditions that allow the application of MLSE in the efficient form of a Viterbi equaliser are revisited. Then, the third section shows simulation results based on the model of the MLSE introduced. The requirements on the clock recovery of an MLSE receiver are discussed. A comparison of the modulation formats CNRZ and ODB modulation in combination with a distortion tolerant MLSE receiver is made.

2.   MLSE FOR NONLINEAR CHANNELS

Maximum likelihood sequence estimation for nonlinear channels relies on a shift register type model with finite memory, i.e., the output signal $y(t)$ during the time interval $(iT, (i+1)T)$ depends only on a finite number of input symbols $\sigma_i = (x_i, x_{i-1}, \ldots, x_{i-m})$, with σ_i denoting the state of the shift register at time instance i

$$y(t) = \sum_i h(t - iT, x_i, x_{i-1}, \ldots, x_{i-m}) + n(t),$$

$$= \sum_i h(t - iT, x_i, \sigma_i) + n(t).$$

The waveform $h(t, x_i, \sigma_i)$, called chip, is zero outside the interval $(0, T)$. The noise process $n(t)$ is assumed to be white and signal independent. The optimum receiver for such a nonlinear channel with finite memory m is a bank of 2^{m+1} filters matched to $h(t, x_i, \sigma_i)$ with symbol spaced sampling and followed by a comparison algorithm selecting the most probable sequence [5],

$$\hat{x} = \arg \max_x \sum_i \log p(y_i(x_i, \sigma_i)|x),$$

where $y_i(x_i, \sigma_i)$ denotes the sampled output at time instance i of the matched filter for $h(t, x_i, \sigma_i)$. Due to the finite memory of the channel, finding the most likely transmitted sequence can be done very efficiently by means of the Viterbi algorithm (see e.g. [6]).

As a matched filter bank is too complex for an optical receiver, a simple low pass filter is used instead for the rest of this paper. The output signal of the low pass filter is sampled with a frequency of s/T resulting in s-fold oversampling for each bit. The branch metric of the Viterbi algorithm finally is approximated as (see e.g. [7])

$$\log p(y_i|\boldsymbol{x}) \approx \sum_{k=1}^{s} \log p(^k y_i|\boldsymbol{x}),$$

where $^k y_i$ denotes the kth sample of symbol i after low pass filtering.

For calculating the branch metrics of the Viterbi algorithm the probability density functions of the sampling values have to be calculated or estimated. In this paper an histogram based method employing a finite quantisation of the sampling values, as suggested in [8], is used for this estimation. This approach automatically accounts for non-Gaussian distribution functions and for signal dependent noise.

Note that the well known union bound approximation for estimating the performance of MLSE for low bit error rates is also valid for finite states non-linear channels as long as the noise is signal independent and its probability distribution function is known. But in contrast to the linear case, the distance of all pairs of sequences have to be considered, not just the distance of all sequences with respect to the all zero sequence. An efficient method to calculate these distances, based on a modification of the Viterbi algorithm, is given in [5]. In case of signal dependent noise as for optically amplified systems with dominating ASE noise, however, the distance of the sequences is not the only metric that determines the error rate of the system. Therefore the union bound approach is no longer valid in its simple form.

3. TRANSMISSION SCENARIO AND RESULTS

The simulation results given in this section focus on a 10.7 GHz metro transmission over standard single mode fibre (SSMF) with a typical dispersion parameter of 17 ps/(nm·km). The transmission model is a single channel, single span scenario with a launch power in the linear regime. The optically preamplified receiver front-end consists of an EDFA and a 50 GHz super-Gaussian filter followed by a photodiode and a 7.5 GHz 5th order Bessel filter. Only optical amplifier noise is considered.

Albeit the MLSE is in principle optimum with respect to the sequence error probability, a feasible implementation has to assume a certain channel memory and to restrict the resolution of the analogue/digital converter. Our simulation model is based on a moderate 4 state trellis and a linear signal quantisation of 4 bits with equally spaced double sampling. For the FFE (DFE) a tapped delay line structure with 6 taps (4 feed forward and 2 feedback taps) is assumed, all of them spaced by $T/2$, where T is the bit time.

The system performance is evaluated by means of Monte-Carlo simulations. It is expressed as required optical signal to noise ratio (OSNR) in 0.1 nm bandwidth to reach a bit error ratio (BER) of $5 \cdot 10^{-4}$, sufficient for error-free operation with enhanced forward error correction (EFEC). For all results an optimized static sampling phase is assumed.

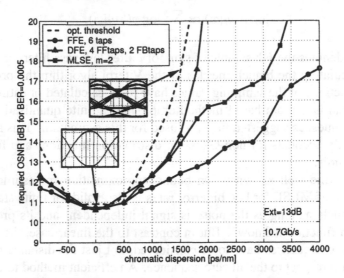

Figure 1.   Required OSNR for a bit error ratio of $5 \cdot 10^{-4}$ vs. chromatic dispersion for standard NRZ transmission and different dispersion compensating receivers.

Fig. 1 displays the required OSNR for increasing chromatic dispersion of NRZ transmission with different receivers. The label "opt. threshold" denotes a standard receiver with threshold optimization. In addition to the MLSE, an FFE and a DFE is shown for comparison. As expected, the MLSE provides the best performance, followed by the DFE. The FFE gives satisfactory results only for low OSNR penalty values, where all equalisers perform very similar.

3.1   Static sampling phase sensitivity

Electrical signal processing not just simply extends the dispersion tolerance, it also influences the sensitivity of the receiver in terms of static sampling phase deviations, as shown in Fig. 2. Given is the required OSNR for a certain bit error ratio for different values of the static sampling phase. For the same value of chromatic dispersion, the sampling phase of an MLSE receiver is less critical compared to a standard receiver with threshold optimization. For increasing values of dispersion, however, the performance of the MLSE gets mutch more sensible to static deviations from the optimum sampling phase. Therefore, in a hardware realisation it is very important that the clock recovery operates reliable at high distortions and a very low BER. The curves for the MLSE are given for half of the sampling phase range only, due to an equally spaced 2-fold oversampling. Higher values of the sampling phase lead to a sampling of the

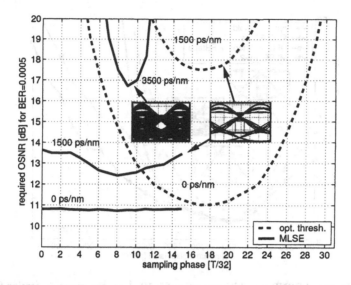

Figure 2.   Sampling phase sensitivity of NRZ for a standard receiver with threshold optimization compared to a MLSE receiver. The required OSNR for a bit error ratio of $5 \cdot 10^{-4}$ is plotted vs. different static shifts of the sampling time. At 3500 ps/nm the eye is completely closed.

neighbouring bit with the second sample and thus to a considerably increased error rate. Note that in case of MLSE reception the optimum sampling phase for 1500 ps/nm and 3500 ps/nm is not at the center of the eye but rather at approx. $T/4$ for the first sample and $3T/4$ for the second one.

3.2   Chirped NRZ modulation

The simulation model for chirped NRZ (CNRZ) transmission assumes a LiNbO$_3$ Mach-Zehnder modulator (MZM) with a fixed chirp parameter (Henry factor) [9] of $\alpha_H = -0.7$. Such a Henry factor can be achieved without additional costs by an asymmetric structure of the modulator arms (as obtained by using z-cut LiNbO$_3$).

The dispersion tolerance of CNRZ is given in Fig. 3. Compared to ordinary NRZ the dispersion tolerance curve of the chirped version is no longer symmetrical with its optimum at zero dispersion but is rather shifted to positive values of the chromatic dispersion with an improvement of approximately 0.5 dB at the optimum value of 600 ps/nm. For both modulation formats, an MLSE allows more than 1000 ps/nm of increased chromatic dispersion at an OSNR penalty of 2 dB. However, whilst the increased dispersion tolerance of NRZ also applies for negative values of the cromatic dispersion, with CNRZ the benefit of the MLSE for negative chromatic dispersion is very small. But this does not matter in the scenario considered in this paper, where no opti-

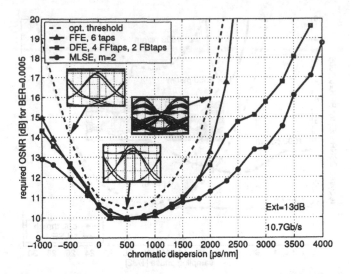

Figure 3. Required OSNR for a bit error ratio of $5 \cdot 10^{-4}$ vs. chromatic dispersion for chirped NRZ transmission with a chirp parameter of $\alpha_H = -0.7$.

cal dispersion compensation is assumed and therefore negative values of the chromatic dispersion cannot occur.

3.3   Optical duobinary modulation

The optical duobinary signal is generated by tight electrical filtering of the Mach-Zehnder driving signal using a 10th order Bessel filter with a FWHM bandwidth of 2.5 GHz. This signal drives the MZM in push-pull configuration with the bias point at zero output, and a driving amplitude of $2V_\pi$, thus creating an optical pseudoternary signal where marks that are separated by an odd number of spaces show a phase difference of π of the electrical field [10]. The optical bandwidth (e.g. from multiplexers) is assumed to be large enough to have no significant influence on the signal.

The results for optical duobinary transmission with different distortion equalizing receivers is given in Fig. 4. At an OSNR penalty of 2 dB the MLSE still can achieve an improvement of approximately 1000 ps/nm in chromatic dispersion tolerance, whereas the gain of the FFE and DFE based equalizers is quite small. Optical duobinary transmission combined with a MLSE receiver is able to reach 250 km of standard single mode fibre (SSMF) without additional dispersion compensation, whereas for chirped NRZ transmission with MLSE only 150 km can be reached. However, the better performance of ODB/MLSE comes at the expense of a more complex and costly transmitter, whereas chirped transmission does not incur any additional cost.

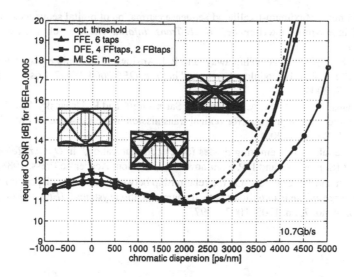

Figure 4. Required OSNR for a bit error ratio of $5 \cdot 10^{-4}$ vs. chromatic dispersion for optical duobinary transmission.

4. CONCLUSIONS

Combining EDC with modulation formats such as chirped NRZ and optical duobinary transmission is a powerful tool to increase the transmission reach of 10 Gb/s metro systems without the need of optical dispersion compensation making it attractive for a seamless upgrade of 2.5 Gb/s systems. Comparing different EDC schemes, MLSE delivers the best performance and allows for ≈ 150 km transmission when combined with CNRZ and ≈ 250 km transmission when combined with ODB. In addition, the use of EDC provides further benefits such as an increased robustness against PMD which cannot be mitigated by CNRZ or ODB modulation.

REFERENCES

[1] J. H. Winters and R.D. Gitlin, "Electrical signal processing techniques in long-haul fiber-optic systems," *IEEE Trans. Commun.*, vol. 38, no. 9, pp. 1439–1453, Sept. 1990.

[2] J. C. Cartledge and R. G. McKay, "Performance of 10 Gb/s lightwave systems using an adustable chirp optical modulator and linear equalization," *IEEE Photon. Tech. Lett.*, vol. 4, pp. 1394–1397, 1992.

[3] G. S. Kanter, A. K. Samal, and A. Gandhi, "Electronic dispersion compensation for extended reach," in *Proc. OFC 2004*, TuG1, Los Angeles, USA, Feb. 22–27, 2004.

[4] O. E. Agazzi and V. Gopinathan, "The impact of nonlinearity on electronic dispersion compensation of optical channels," in *Proc. OFC 2004*, TuG6, Los Angeles, USA, Feb. 22–27, 2004.

[5] S. Benedetto, E. Biglieri, and V. Castellani, *Digital Transmission Theory*, Prentice Hall, 1987.

[6] D. G. Forney, "Maximum-likelihood sequence estimation of digital sequences in the presence of intersymbol interference," *IEEE Trans. Inform. Theory*, vol. 18, pp. 363–378, May 1972.

[7] C. R. S. Fludger, J. E. A. Whiteaway, and P. J. Anslow, "Electronic equalisation for low cost 10 Gbit/s directly modulated systems," in *Proc. OFC 2004*, WM7, Los Angeles, USA, Feb. 22–27, 2004.

[8] H. F. Haunstein, K. Sticht, A. Dittrich, W. Sauer-Greff, and R. Urbansky, "Design of near optimum electrical equalizers for optical transmission in the presence of PMD," in *Proc. OFC 2001*, WAA4-1, 2001.

[9] T. P. Lee, et al., "Light emitting diodes for telecommunication," in *Optical Telecommunications II* (S. E. Miller and I. P. Kaminow, eds.), London: Academic Press, 1988.

[10] S. Walklin and Jan Conradi, "On the relationship between chromatic dispersion and transmitter filter response in duobinary optical communication systems," *IEEE Photon. Technol. Lett.*, vol. 9, no. 7, pp. 1005–1007, 1997.

Author Index

Index

216